Aircraft Modeling and Simulation

Aircraft Modeling and Simulation

Editors

Ruxandra Mihaela Botez
Teodor Lucian Grigorie
Oliviu Sugar-Gabor
Alejandro Murrieta-Mendoza

MDPI • Basel • Beijing • Wuhan • Barcelona • Belgrade • Manchester • Tokyo • Cluj • Tianjin

Editors

Ruxandra Mihaela Botez
École de technologie supérieure
ÉTS, Université du Québec
Canada

Teodor Lucian Grigorie
Military Technical Academy
"Ferdinand I"
Romania

Oliviu Sugar-Gabor
University of Salford
UK

Alejandro Murrieta-Mendoza
Amsterdam University of
Applied Science
The Netherlands

Editorial Office
MDPI
St. Alban-Anlage 66
4052 Basel, Switzerland

This is a reprint of articles from the Special Issue published online in the open access journal *Applied Sciences* (ISSN 2076-3417) (available at: https://www.mdpi.com/journal/applsci/special_issues/Aircraft_Modeling_Simulation).

For citation purposes, cite each article independently as indicated on the article page online and as indicated below:

LastName, A.A.; LastName, B.B.; LastName, C.C. Article Title. *Journal Name* **Year**, *Volume Number*, Page Range.

ISBN 978-3-0365-3210-3 (Hbk)
ISBN 978-3-0365-3211-0 (PDF)

Cover image courtesy of Ruxandra Mihaela Botez

Contents

About the Editors

Ruxandra Mihaela Botez obtained her degree in Aerospace Engineering from the Faculty of Aerospace Design in Bucharest, Romania, in 1984. Ruxandra obtained her Master's in Applied Sciences degree from École Polytechnique in 1989 and her PhD degree from McGill University in Montreal, Canada, in 1994. She worked as a Postdoctoral Fellow at Auburn University until 1995. Dr Botez has Aerospace Engineering experience acquired in two aerospace companies: IAR-Brasov in Romania (1984–1987) and Bombardier Aerospace in Montreal, Canada (1995–1997). Dr Ruxandra Mihaela Botez has been a Full Professor and the Head of the Laboratory of Applied Research in Active Controls, Avionics, and AeroServoElasticity LARCASE at the École de technologie supérieure ÉTS in Montreal, Canada, since 1998. Dr Botez has been a Canada Research Chair Tier 1 Holder in Aircraft Modeling and Simulation Technologies since 2011. Through her research projects, Dr Botez advised more than 264 Bachelor's internship students, 128 Master's in Engineering students, and 20 PhD Students. She is an Associate Fellow of the American Institute of Aeronautics and Astronautics (AIAA) and of the Canadian Aeronautical Society Institute (CASI) and a Fellow of the Royal Aeronautical Society (RAeS) and of the International Association of Advanced Materials (IAAM). Dr Botez and her team work in collaboration with aerospace companies, research institutes and universities in Canada (Bombardier Aerospace, Bell Helicopter Textron, CAE Inc., CMC Electronics-Esterline, IAR-NRC), in Italy (University of Naples, University of Bologna and CIRA), in Romania (INCAS, Military Technical Academy), in Mexico (Hydra Technologies), in Brazil (UNESP), in the USA (NASA), and in other countries. Dr Botez is Editor-in-Chief of the INCAS Bulletin, edited by the National Institute for Aerospace Research Elie Carafoli INCAS in Romania. She and her team have published more than 500 scientific publications, including 160 journal papers, 300 conference proceedings papers, and 7 book chapters; Dr Botez has also written six textbooks for teaching in coureses and she is the co-editor of one book. Dr Botez and her team have won more than 50 awards, and their research is mentioned in more than 100 media articles.

Teodor Lucian Grigorie was born in 1975 in Craiova, Romania, where he received a BsC and an MsC in Aerospace Engineering from the Faculty of Electrical Engineering, University of Craiova, in 1998 and 1999, respectively. In 2006 he obtained a PhD at the Faculty of Aerospace Engineering, Politehnica University of Bucharest, in the field of Aerospace Engineering. From 1998 to 2005 he worked as an Assistant Professor in the Avionics Division of University of Craiova, and from 2005 to 2009, he was Senior Lecturer in inertial navigation systems at the same institution. From 2009 to 2018 he was an Associate Professor in the Avionics Division of the University of Craiova. Between May 2008 and February 2009, he was a postdoctoral fellow at the University of Quebec, Canada, an institution with whom he has collaborates with on research projects since 2002. Between October 2018 and February 2020, he was a First Degree Scientific Researcher at Military Technical Academy "Ferdinand I" in Bucharest, Faculty of Aircraft and Military Vehicles, Center of Excellence in Self-Propelled Systems and Technologies for Defense and Security, and Since February 2020, he has been a Full Professor at the same institution in the Department of Aircraft Integrated Systems and Mechanics. He has been a PhD supervisor at University Politehnica of Bucharest, Faculty of Aerospace Engineering, since October 2017. He has authored or co-authored 5 books, 2 laboratory handbooks, 6 book chapters, and more than 200 papers in journals and conference proceedings. His research activities concern aerospace navigation systems, miniaturized inertial sensors, avionics equipment and systems, automatic control systems, electric actuation systems (including smart material actuators), and morphing aircraft structures.

Oliviu Sugar-Gabor obtained his BEng degree in Aerospace Engineering from the Faculty of Aerospace Engineering, Politehnica University of Bucharest, in 2009, and his MSc in Aerospace Engineering (Numerical Simulation Specialization) from the same institution in 2011. In 2015 he received his PhD from the École de technologie supérieure (ÉTS) in Montreal, Canada. Since 2016, he has been a Lecturer (Assistant Professor) in aerodynamics at the University of Salford, Manchester, in the United Kingdom. His research interests include applied aerodynamics, computational fluid dynamics, aerodynamic tools for aircraft design, non-intrusive reduced order models, and morphing wing concepts. He is a full Member of the Royal Aeronautical Society (RAeS) and of the Higher Education Academy (HEA). He has authored and co-authored 47 papers in journals and conference proceedings, 1 book chapter, as well as 1 laboratory handbook and 2 textbooks for undergraduate and postgraduate teaching.

Alejandro Murrieta-Mendoza obtained his BEng in Electronics Engineering from the Autonomous University of Baja California (UABC) in Mexico in 2007. He obtained his M. Eng in Aerospace Engineering in 2013 and his Ph.D in Aerospace Engineering in 2017 from L'École de Technologie Supérieure (ÉTS). He has been a Lecturer/Researcher at the Aviation Academy at the Amsterdam University of Applied Sciences in the Netherlands since 2017. He has published 11 journal articles, presented over 25 conference papers, and published a book chapter. Dr. Murrieta-Mendoza has been awarded the 2020 SAE® Charles M. Manly Memorial Medal and has been the recipient of excellence scholarships from the governments of Mexico via CONACyT and the government of Quebec via the FRQNT. He has been a guest editor for two Special Issues covering trajectory optimization at the Aerospace journal by MDPI. His research covers the applications of simulation, optimization techniques, machine learning, and artificial intelligence applied to aviation and logistics problems and applications.

applied sciences

Editorial

Editorial for the Special Issue "Aircraft Modeling and Simulation"

Ruxandra Mihaela Botez

Laboratory of Applied Research in Active Controls, Avionics and AeroServoElasticity LARCASE, ÉTS—École de Technologie Supérieure, Université de Québec, 1100 Notre-Dame Street West, Montréal, QC H3C 1K3, Canada; ruxandra.botez@etsmtl.ca; Tel.: +1-514-396-8560

1. Introduction

New airplane and unmanned aerial system modeling, simulation, and design methodologies are very important in aerospace engineering. The best methodologies should be selected in order to reduce the need for a high number of expensive experimental data (and, thus, to minimize fuel consumption). These methodologies should be applied on an aircraft with the aim of certifying it for production. Experimental data are usually provided by use of wind tunnel and flight tests.

Therefore, in this Special Issue, numerical methodologies for computational fluid dynamics (CFD), structural dynamics, and controls were studied, and their results were compared and validated with experimental wind tunnel data and other numerical methodologies' results.

The aim of this Special Issue was to promote research and development on aircraft modeling and simulation technologies, while addressing their validation with a minimum possible amount of experimental data.

2. Summary of the Special Issue Contents

The Special Issue can be divided into three groups according to the disciplines and their applications on systems, covered by five articles. These disciplines are focused on aerodynamics, structural dynamics, and controls.

In the first group, focused on the aerodynamics or computational fluid dynamics (CFD) areas or research, two articles were published on the following: the first covered aerodynamics analyses applications on the droop nose leading edge (DNLE) with the morphing trailing edge (MTE) combination for the UAS-S45 morphing wing [1]; the second was focused on the following four different configurations: a flat plate, an airfoil near-wake, a backward-facing step, and a turbine cascade, also called the eleventh standard configuration [2]. These aerodynamic studies were performed using various CFD in-house and commercial software, and their results were validated with experimental data provided in the literature.

In the second group, focused on structural dynamics modeling area, two articles were published on an adaptive winglet finite element model (FEM) issue [3] and on the modeling of an oleo-pneumatic landing gear using MATLAB instead of FEM, which was considered as one of the originalities of this paper [4].

In the third area, focused on controls, one paper was written on the design and wind tunnel test validation of a disturbance rejection dynamic inverse control for a tailless aircraft [5].

2.1. Study Case–Unmanned Aerial System UAS-S45 Morphing Wing Aerodynamic Analysis

Among green aircraft technologies, one might include morphing aircraft systems development. Morphing or adaptive wing and winglets are able to change their structural

Citation: Botez, R.M. Editorial for the Special Issue "Aircraft Modeling and Simulation". *Appl. Sci.* **2022**, *12*, 1234. https://doi.org/10.3390/app12031234

Received: 16 January 2022
Accepted: 21 January 2022
Published: 25 January 2022

Publisher's Note: MDPI stays neutral with regard to jurisdictional claims in published maps and institutional affiliations.

shapes using actuators, sensors, and controls technologies in order to obtain better aerodynamic performances for the aircraft, as shown in the aerodynamics studies for morphing wings [1] and in the structural studies of an adaptive winglet [3].

In [1], an aerodynamic optimization new methodology was employed for a combination of the droop nose leading edge (DNLE) with the morphing trailing edge (MTE) of an UAS-S45 root airfoil by use of the Bezier-PARSEC parameterization. This methodology used a hybrid optimization technique, based on a combination of the particle swarm optimization (PSO) and pattern search algorithms.

The drag was minimized and the endurance maximized for the UAS-S45. The aerodynamic analysis results were obtained for the UAS-S45 airfoil using the XFoil software, and for the UAS-S45 wing using the high-fidelity computational fluid dynamics (CFD) Ansys Fluent solver including the transition ($\gamma - Re\theta$) shear stress transport (SST) turbulence model. The aerodynamic optimization results were obtained for different flight conditions. Both the DNLE and MTE optimized airfoils have shown a significant improvement in the UAS-S45 overall aerodynamic performance, while the MTE airfoils increased the efficiency of $C_L^{3/2}/C_D$ by 10.25%, thus indicating better endurance performance. Therefore, both DNLE and MTE configurations have shown promising results in improving the aerodynamic efficiency of the UAS-S45 airfoil.

2.2. Study Case–Comparison between CFD and Experimental Results Using Three Different Software

In the aerospace industry, computational fluid dynamics (CFD) methodologies are researched for advancing aerodynamics studies on aircrafts. The results obtained by these methodologies are compared among themselves, and with experimental wind tunnel tests and flight tests results. The laminar, turbulent, and transition flow results are analyzed using these numerical methodologies.

In this study case [2], two turbulence models—the shear stress transport (SST) model and the Spalart–Allmaras (SA) model—were implemented in the UNS3D in-house code at the Texas A&M University, and their results were compared with those of FUN3D and CFL3D codes developed by NASA. The UNS3D code has two versions: UNS3D-SEQ (sequential version) and UNS3D-PAR (parallel version). In addition, these numerical results were compared with experimental results from the literature. The methodologies were applied on four different configurations: a flat plate, an airfoil near-wake, a backward-facing step, and a turbine cascade, also called the eleventh standard configuration.

Regarding the comparison of the results, the solutions' residuals were very small, more precisely, less than 10^{-11}. The SST model predicted, better than the SA model, the turbulent fluctuations and skin friction coefficients in comparison with experimental data, while the SA model predicted better than the SST model, as the flow went away from the backward-facing step. In fact, most of the results obtained using the SST model fitted the experimental data better than the SA model, while the main disadvantage of the SST model resided in its computational execution time, that was higher than the SA model execution time, with its values between 4–38%.

2.3. Study Case–Structural Analysis of an Adaptive Winglet

Adaptive and morphing surfaces of aircraft are studied worldwide with the aim of improving aerodynamic performance. Among these surfaces, winglets are often studied. At CIRA, in Italy, the structural team has been continuously working in this interesting area.

The finite element modeling (FEM) issues for an adaptive winglet skeleton design at CIRA are discussed by [3]. For example, in this paper, a study was presented on the structural architecture adaptation for a winglet morphing system in order to allow its deformations within the safety margins. Regarding this structural morphing winglet design, FEM solver problems occurred as the safety factors (including those for severe load conditions) were highly dependent on the mesh sizes. As the mesh was refined, the singularities were represented through single points or lines. This study was focused mainly on the presentation of causes and their effects on the results. In addition, some experimental issues were also discussed in this paper regarding the adaptive winglet skeleton.

2.4. Study Case–Oleo-Pneumatic Landing Gear System Drop Impact Dynamics

Oleo-pneumatic landing gear is a complex component and system that is usually designed in parallel with other components of an aircraft, such as the fuselage and wings. FEM is usually employed for modeling and analyzing this system, which might have a high impact on the structural aircraft dynamics.

In [4], a new methodology is shown, in which four state variables are considered for the modeling and simulation of the oleo-pneumatic landing gear drop impact dynamics. The forces obtained during the drop were simulated on both horizontal and vertical axes. The well-known MATLAB software considered a set of intercommunicating routines, and it was used instead of FEM software for modeling and simulating the drop impact dynamics, and the numerical results were validated with experimental data. The advantages and limitations of these studies were discussed in this paper.

2.5. Study Case–Analysis and Wind Tunnel Tests for the Nonlinear Dynamic Inversion (NDI) Control Methodology on a Tailless Aircraft

In [5], the design and wind tunnel tests of the validation of a disturbance rejection dynamic inverse control for a tailless aircraft are presented. In this paper, nonlinear dynamic inversion (NDI)-based disturbance rejection control methodologies were designed, and then wind tunnel tests were used to validate the numerical methodologies' results.

A nonlinear affine mathematical model was obtained numerically for the tailless aircraft model supported with a 3-DOF rig in the wind tunnel. A baseline NDI controller was designed to stabilize and control the aircraft attitude; this controller was further augmented with a disturbance observer, and became an NDI-DO controller, that was used to reject the lumped disturbances. The simulation has shown that the robustness of the NDI-DO augmented controller was higher than the robustness of the baseline NDI controller, and that the anti-windup (AW), modified disturbance observer recovered the control performance from the actuator saturation.

Finally, wind tunnel tests were successfully conducted, and their experimental results validated the simulation results obtained by the NDI-DO control methodology; thus, the experimental results fitted the simulation results using the NDI-DO controller very well, which demonstrated a higher tracking and more robust performance than the NDI-PI (proportional integral) controller. However, the NDI-AW controller was not implemented and tested due to the absence of sensors for the actual surface deflections.

3. Conclusions

As seen in this Special Issue, "Aircraft Modeling and Simulation", aerodynamics, structures, and controls engineering analyses and experimental tests were presented for different aircraft systems and configurations for aircraft and unmanned aerial systems. Morphing aerodynamic and structural analyses studies were presented for a morphing wing of the UAS-S45 from Hydra Technologies in Mexico, and for an adaptive structural winglet study with the aim to advance green aircraft technologies, by improving aerodynamic performance in terms of fuel consumption and greenhouse emissions reduction. A deep CFD analysis study using an in-house developed software UNS3D code was also performed for four configurations, and its results were compared with other NASA software results and experimental data. A structural analysis of an oleo-pneumatic landing gear was presented by use of MATLAB instead of classical FEM analysis, while a control analysis was numerically and experimentally tested in the wind tunnel for a tailless aircraft. Finally, these studies are extremely important in the advancement of aircraft engineering research.

Funding: This research was supported from the Canada Research Chair in Aircraft Modeling and Simulation funds (nr 231679).

Acknowledgments: This publication was only possible with the valuable contributions from the authors, reviewers, and the editorial team of *Applied Science*. In particular, we would like to thank Frederic Yuan, for his tireless and efficient support.

Conflicts of Interest: The author declares no conflict of interest.

References

1. Bashir, M.; Longtin-Martel, S.; Botez, R.; Wong, T. Aerodynamic Design Optimization of a Morphing Leading Edge and Trailing Edge Airfoil–Application on the UAS-S45. *Appl. Sci.* **2021**, *11*, 1664. [CrossRef]
2. Polewski, M.; Cizmas, P. Several Cases for the Validation of Turbulence Models Implementation. *Appl. Sci.* **2021**, *11*, 3377. [CrossRef]
3. Ameduri, S.; Dimino, I.; Concilio, A.; Mercurio, U.; Pellone, L. Specific Modeling Issues on an Adaptive Winglet Skeleton. *Appl. Sci.* **2021**, *11*, 3565. [CrossRef]
4. Pecora, R. A Rational Numerical Method for Simulation of Drop-Impact Dynamics of Oleo-Pneumatic Landing Gear. *Appl. Sci.* **2021**, *11*, 4136. [CrossRef]
5. Nie, B.; Liu, Z.; Guo, T.; Fan, L.; Ma, H.; Sename, O. Design and Validation of Disturbance Rejection Dynamic Inverse Control for a Tailless Aircraft in Wind Tunnel. *Appl. Sci.* **2021**, *11*, 1407. [CrossRef]

applied
sciences

MDPI

Aerodynamic Design Optimization of a Morphing Leading Edge and Trailing Edge Airfoil–Application on the UAS-S45

Musavir Bashir, Simon Longtin-Martel, Ruxandra Mihaela Botez * and Tony Wong

Research Laboratory in Active Controls, Avionics and Aeroservoelasticity (LARCASE), Université du Québec, École de Technolgie Supérieure, 1100 Notre-Dame West, Montreal, QC H3C1K3, Canada; musavir-bashir.musavir-bashir.1@ens.etsmtl.ca (M.B.); simon.longtin-martel.1@ens.etsmtl.ca (S.L.-M.); tony.wong@etsmtl.ca (T.W.)
* Correspondence: ruxandra.botez@etsmtl.ca

Abstract: This work presents an aerodynamic optimization method for a Droop Nose Leading Edge (DNLE) and Morphing Trailing Edge (MTE) of a UAS-S45 root airfoil by using Bezier-PARSEC parameterization. The method is performed using a hybrid optimization technique based on a Particle Swarm Optimization (PSO) algorithm combined with a Pattern Search algorithm. This is needed to provide an efficient exploitation of the potential configurations obtained by the PSO algorithm. The drag minimization and the endurance maximization were investigated for these configurations individually as two single-objective optimization functions. The aerodynamic calculations in the optimization framework were performed using the XFOIL solver with flow transition estimation criteria, and these results were next validated with a Computational Fluid Dynamics solver using the Transition $(\gamma - Re_\theta)$ Shear Stress Transport (SST) turbulence model. The optimization was conducted at different flight conditions. Both the DNLE and MTE optimized airfoils showed a significant improvement in the overall aerodynamic performance, and MTE airfoils increased the efficiency of $C_L{}^{3/2}/C_D$ by 10.25%, indicating better endurance performance. Therefore, both DNLE and MTE configurations show promising results in enhancing the aerodynamic efficiency of the UAS-S45 airfoil.

Keywords: morphing airfoil optimization; parameterization; PSO; aerodynamic performance

Citation: Bashir, M.; Longtin-Martel, S.; Botez, R.M.; Wong, T. Aerodynamic Design Optimization of a Morphing Leading Edge and Trailing Edge Airfoil–Application on the UAS-S45. *Appl. Sci.* **2021**, *11*, 1664. https://doi.org/10.3390/app11041664

Academic Editor: Kambiz Vafai

Received: 10 December 2020
Accepted: 6 February 2021
Published: 12 February 2021

Publisher's Note: MDPI stays neutral with regard to jurisdictional claims in published maps and institutional affiliations.

1. Introduction

1.1. Outline of the Research

The goal of reducing global fuel consumption and fuel-related emissions has placed tremendous pressure on the aviation industry. In 2018, the aviation industry was responsible for 895 million tons of CO_2 emissions globally emitted into the atmosphere [1]. Reducing fuel consumption will benefit both the world environment and the air transport industry. According to the Aviation Transport Action Group (ATAG), a reduction in fuel burning will have a significant impact on the aviation industry because the largest operating cost of this industry is fuel [1]. Drastic measures are still required even though some steps have already been taken to reduce these emissions. Various solutions have been implemented by the aviation industry, including smart material technology, laminar flow technology, air traffic management technologies, advanced propulsion techniques and sustainable fuels [2–4]. Morphing wing technology is one of the technologies showing high potential in decreasing aircraft fuel consumption [5–8]. Even though there is no settled definition for "morphing", this term is borrowed in Aviation Technology from avian flight to describe the ability to modify maneuvers at certain flight characteristics in order to obtain the best possible performance. This type of application requires morphing structures capable of adapting to changing flight conditions. Morphing systems include several wing shape modifications, span or sweep changes, changes in twist or dihedral angle [9] and variations of the airfoil camber [7] or of the thickness distribution [4]. In the design phase, both military and civilian aircraft traditionally fly at a single or few optimum flight conditions; morphing, however, is envisioned to increase the number of optimum operational points for a given aircraft [5].

The morphing strategy scope is broad in the sense that an optimal solution can be generated in various ways with respect to a wide spectrum of mission profiles and types, and flight regimes. Novel technologies need to be part of recent efforts to produce environmentally sustainable and efficient aircraft. To meet environmental guidelines, including those of the International Civil Aviation Organization's (ICAO's) Committee on Aviation Environmental Protection (CAEP), innovative technologies such as "morphing" are fascinating as they provide advantages over conventional wing configurations [10]. Additionally, "morphing" is more practically applied in Unmanned Aerial Vehicles (UAVs) because of their reduced scale and lower complexity in terms of wing design structure and energy consumption expressed in terms of actuation power [6,11,12], in addition to the advantages that morphing designs are lighter and less noisy than their conventional designs. Some morphing opportunities with potential benefits for increasing aerodynamic performance of the UAS-S45 are shown in Figure 1.

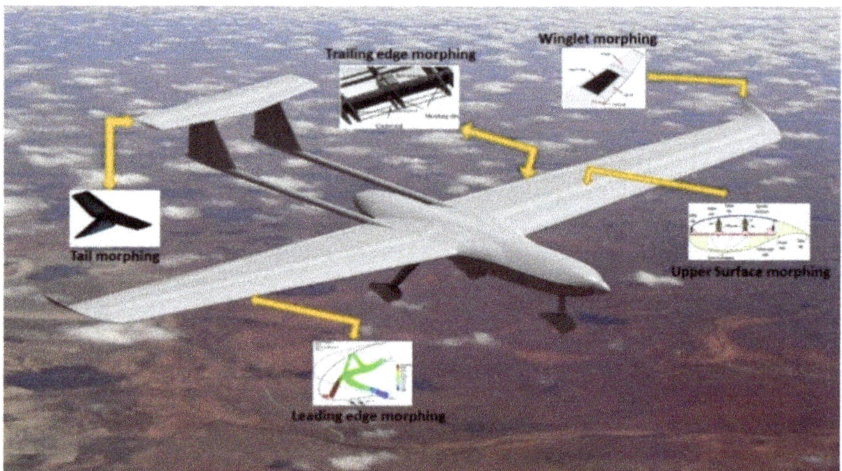

Figure 1. UAS-S45 with Potential Morphing Configuration Capabilities.

Aircraft are designed to achieve optimal aerodynamic performance (maximum lift-to-drag ratio) for a mission specification and for an extended range of flight conditions. Nevertheless, these mission specifications change continuously throughout the different flight phases, and an aircraft often flies at non-optimal flight conditions. Although the conventional hinged high lifting devices and trailing edge surfaces (discrete control surfaces) in aircraft are effective in controlling the airflow for different flight conditions, they create surface discontinuities, which increase the drag [13,14]. The disadvantages of these hinged surfaces are found in both their deployed and retracted configurations [15]. When deployed, the gaps between the high lifting surface and the wing can cause noise and turbulence, and therefore can produce a turbulent boundary layer and thus increase drag. Even when retracted, the trailing edge hinges still produce a turbulent boundary layer. Numerous researchers believe that the laminar flow technology has the maximum potential to reduce drag and to avoid flow separation [16,17]. For this goal, the wings require the design of thin airfoils, seamless high-quality surfaces and variable droop leading edges. The use of these high continuous smooth lifting devices in UAVs will further broaden their flight envelope and extend their endurance. These challenges can be addressed by using morphing technology in the current aviation industry.

The aim of this study is thus to investigate the aerodynamic optimization by employing constrained shape parameterization for UAS-S45 root airfoil using Droop Nose Leading Edge (DNLE) and Morphing Trailing Edge (MTE) technologies. The aerodynamic design of continuous morphing wing surfaces employed to replace traditional discrete control surfaces, such as flaps, ailerons or sometimes slats, to adjust the camber of the wing, are discussed in this paper along with the parameterization strategy and the optimization algorithm.

1.2. Literature Survey

The potential in the field of aircraft morphing is evident as hundreds of research groups worldwide have been dedicated to studying the different aspects of morphing. Two of the earliest groups were launched by NASA, and they worked within the "Active Flexible Wing program" and the "Mission Adaptive Wing program" [18,19]. In these programs, flutter suppression, load alleviation and load control during rapid roll maneuvers were investigated.

One of the potential consequences of flexible morphing structures is that both dynamic properties and aerodynamic loads of the wing are affected. Hence, the aeroelastic effects arising from interactions between the configuration-varying aerodynamics and the morphing structure are significant. Li et al. have presented important morphing studies in terms of aeroelastic, control and optimization aspects [20].

The nonlinear aeroelastic behaviour of a composite wing with morphing trailing-edge has been studied [21]. Moreover, the nonlinear aeroelastic behaviour of a composite wing with morphing trailing-edge has been studied.

Aero-elastically stable configurations of a morphing wing trailing edge driven by electromechanical actuators have been investigated [22]. The effects induced by trailing-edge actuators stiffnesses on the aeroelastic behaviour of the wing were simulated using different approaches. The results revealed that flutter could be avoided if sufficient stiffness was provided by these actuators.

An analytical sensitivity calculation platform for flexible wings has been studied [23,24]. This platform has been used to perform wing aeroelastic optimization and stability analysis.

In another study, a low-fidelity model of an active camber aero-elastic morphing wing was developed to investigate the critical speed values by varying its chord-wise dimension [25]. A wide range of configurations was explored to predict the dynamic behaviour of these active camber morphing wings.

The combined flutter behaviour and gust response of a series of flexible airfoils has been investigated [26]. Rayleigh's beam equation was used to model this series of flexible airfoils. The effect of chordwise flexibility of a compliant airfoil was investigated numerically to demonstrate its dynamical stability [27,28]. An actuated two-dimensional membrane airfoil was investigated experimentally and numerically, and it was found that membrane flexibility decreased the drag and delayed the stall.

These aeroelastic studies are significant in understanding the flexible morphing wing and its potentially relevant effects. However, our study does not consider the structural effects of morphing, and therefore the optimization is aimed to improve the aerodynamic drag performance and endurance in the given range of angles of attack.

Other research projects in the US and Canada include the "Smart Materials and Structures Demonstration" program initiated to develop new affordable smart materials to establish the performance gains by investigating smart rotors, the smart aircraft and the marine propulsion system; the "Aircraft Morphing" program [29]; the "Advanced Fighter Technology Integration" (AFTI) program, aimed to develop and demonstrate in flight a smooth variable camber wing and flight control system capable of adjusting the wing's shape in response to flight conditions to maximize aerodynamic efficiency; the "Active Aeroelastic Wing" program, demonstrated to improve aircraft roll control through aerodynamically induced wing twist on a full-scale high-performance aircraft at transonic and supersonic speeds; the "Morphing Aircraft Structures" program [30]; the

"Mission Optimized Smart Structures" (MOSS), in which aerodynamic optimization was performed with the stretching of the leading edge using a novel skin material developed with a nanocomposite at the National Research Council of Canada [31]; and the "Controller Design and Validation for Laminar Flow Improvement on a Morphing Research Wing–Validation of Numerical Studies with Wind Tunnel Tests—CRIAQ MDO 7.1" [32,33]. The CRIAQ 7.1 project took place at our Laboratory of Applied Research in Active Controls, Avionics and AeroServoElasticity LARCASE.

The European Union has conducted several projects, including the "Active Aeroelastic Aircraft Structures" (3AS) project, to develop novel active aeroelastic control strategies to improve aircraft performance (structural weight, better control effectiveness) by controlling structural deformations to modulate the desired aerodynamic deformations [34]; the "Aircraft Wing Advanced Technology Operations" (AWIATOR) project, in which novel fixed wing configurations were introduced aimed at reducing the vortex hazard by implementing larger winglets and further improving aircraft efficiency and reducing farfield impact [35]; the "New Aircraft Concepts Research" (NACRE) project, in which Powered Tails and Advanced Wings were studied to obtain high environmental performance (noise and CO2 emissions)—both high aspect ratio low-sweep wings and forward-swept wings (with natural laminar flow) contributed to achieving good fuel efficiency [36]; the "Smart Leading Edge Device" (SmartLED) project [13]; the "Smart High Lift Devices for Next Generation Wings" (SADE) project [37]. Both SmartLED and SADE were aimed to develop and investigate the morphing high lift devices, "smart leading edge" to enable seamless high lift devices and therefore enable laminar wings and "smart single-slotted flap", for the next generation aircraft of high surface quality for drag reduction. Projects by the European Union also included the "Smart Fixed Wing Aircraft" (SFWA) project [38] and the "Smart Intelligent Aircraft Structures" (SARISTU) project to address the physical integration of smart intelligent structural concepts. The SARISTU included a series of research collaborations to addresses aircraft weight and operational cost reductions as well as an improvement in the flight profile specific aerodynamic performance [39]. Other projects by the European Union were the "Novel Air Vehicle Configurations" (NOVEMOR) project, in which morphing wing solutions (span and camber strategies and wing-tip devices) were proposed to enhance lift capabilities and maneuvering [40]; the "Clean-Sky 1 and 2" projects [41,42]; the "Combined Morphing Assessment Software Using Flight Envelope Data" (CHANGE) project, which developed a modular software architecture capable of determining and achieving optimum wing shape [43]; and the "Sustainable and Energy Efficient Aviation" (SE2A) project aimed to investigate the Morphing structures for the 1g-wing to exploit the nonlinear structural behavior of wing design components to achieve passive load alleviation [44].

The "Morphing Architectures and Related Technologies for Wing Efficiency Improvement-CRIAQ MDO 505" was also realized at the LARCASE in the continuation of the CRIAQ 7.1 project mentioned above. The achievements of the international Canadian-Italian CRIAQ MODO 505 project are mentioned in various publications [45,46].

The aircraft optimization process has evolved dramatically over the past decades. The design of new intelligent algorithms and computational solvers has substantially impacted the overall design process, including the "morphing aircraft optimization". The verification and validation of aircraft design using optimization techniques has reduced its huge experimental costs and has been achieved by the use of efficient algorithms and computational solvers [47,48]. An optimization process is initiated by minimizing or maximizing an objective function for the target design concept with respect to design variables subjected to given constraints, which were applied to limit the search space and therefore to yield physically feasible optimization results. Both gradient-based and gradient-free intelligent algorithms have been used in these optimization processes, and their results are dependent on the type of optimization problem to be solved [47]. These algorithms are discussed in the following sections. Many optimization techniques based on bioinspired processes and Surrogate-Assisted natural methods have been formulated, and mathematical and statistical analysis of these algorithms has been performed [49–52]. The comparative performance study based on convergence trends was performed with different geometry parameterization techniques. Both advantages and disadvantages can be found in these optimization algorithms for complex shape functions. However, the goal of this study is to perform aerodynamic design optimization of morphing airfoil using Particle Swarm Optimization (PSO) algorithm combined with Pattern Search technique.

Similarly, various computational solvers are employed based on their desired solution accuracy and computational cost [53]. Researchers have conducted various aerodynamic and structural optimization studies, including design studies of different surfaces (components) of the wing, such as its upper surface, trailing edge or leading edge. Adjoint-based airfoil optimization has been studied using Euler equations [54], while other researchers have implemented extensive gradient-based aerodynamic shape optimization methodologies for transonic wing design using high-fidelity RANS solvers [55]. Many other optimization studies have been performed using different optimization techniques. Some of the most notable works are summarized in Table 1.

Table 1. Airfoil optimization research carried out by using different techniques.

Authors	Reference	Category	Optimization Techniques
Gabor, Simon et al., 2016	[56]	Several Morphing Geometries	Artificial Bee Colony Algorithm and a Classical Gradient-Based Search Routine
Lyu, Kenway et al., 2014	[55]	Camber Morphing	Gradient-Based Optimization Algorithm with an Adjoint Method
Hashimoto, Obayashi et al., 2014	[57]	High Wing Configuration	Kriging Surrogate-Assisted Genetic Algorithm
Ganguli and Rajagopal 2009	[58]	Preliminary UAV Design	Kriging Surrogate-Optimization Model
Koreanschi, Gabor et al., 2017	[32]	Wing Tip	Artificial Bee Colony and a Gradient Method
Fincham and Friswell 2015	[59]	Camber Morphing	Pareto Frontiers with a Genetic Algorithm
Murugan, Woods et al., 2015	[60]	Camber Morphing	Pareto Frontiers with a Genetic Algorithm
Albuquerque 2017	[61]	Several Morphing Geometries	Multidisciplinary Design Optimization with Gradient-Based Optimization Algorithm
Khurana, Winarto et al., 2008	[62]	Several UAV Geometries	Particle Swarm Optimization (PSO) Algorithm
Kao, Clark et al., 2019	[63]	Several Morphing Geometries	Gradient-Based Optimization
Magrini, Benini et al., 2019	[64]	Leading Edge Morphing	Multi-Objective Optimization with a Genetic Algorithm
Gong and Ma 2019	[65]	Variable Sweep, Span and Chord Morphing	Genetic Algorithm in conjunction with a Surrogate Model

The study presented in this paper analyses the aerodynamic optimization of the Hydra Technologies S45 Balaam UAS airfoil. This unmanned aerial surveillance (UAS) system is needed to provide security and surveillance capabilities for the Mexican Air Forces, as well as civilian protection in dangerous situations. This study of Morphing Wing Technology has the aim to enhance the aerodynamic efficiency and the effective range of the UAS S45, as well as to extend the flight time. The next section of this paper presents the optimization framework of the overall methodology, which includes two objective functions' formulations ("drag minimization" and "endurance maximization"). The parameterization technique is also presented in Section 2 and has the aim to obtain the "optimal aerodynamic shapes"; the computational solvers for calculating aerodynamic coefficients and the optimization algorithms employed are also presented. The results obtained by these solvers and algorithms for the UAS-S45 optimized designs are compared with the baseline UAS-S45 model results.

2. Method

This optimization framework study performs the aerodynamic optimization of the leading edge and trailing edge of a morphing airfoil by using an intelligent and iterative process based on user-defined aerodynamics and constraints. The methodology involves the objective function formulation integrated with a geometrical shape parameterization model, an aerodynamic flow solver and the optimization algorithm.

For this study, an optimization framework is designed to allow the integration of shape generation using the direct manipulation of the airfoil shape variables while respecting the geometrical constraints. The optimization procedure framework is described in Figure 2, and it consists of a parameterization block based on the Bezier–Parsec (BP) parameterization airfoil shape technique and an aerodynamic solver, such as the panel solver XFOIL (Version 6.99, MIT, USA, 2013). The validation of its results is done using high-fidelity solver in Fluent based on Reynolds-averaged Navier–Stokes (RANS) equations in the Transition SST model. A hybrid optimizer based on Particle Swarm Optimization (PSO) algorithm combined with the Pattern Search technique is also used. The optimization procedure is shown in Figure 2 and is described in detail in Sections 2.1–2.4.

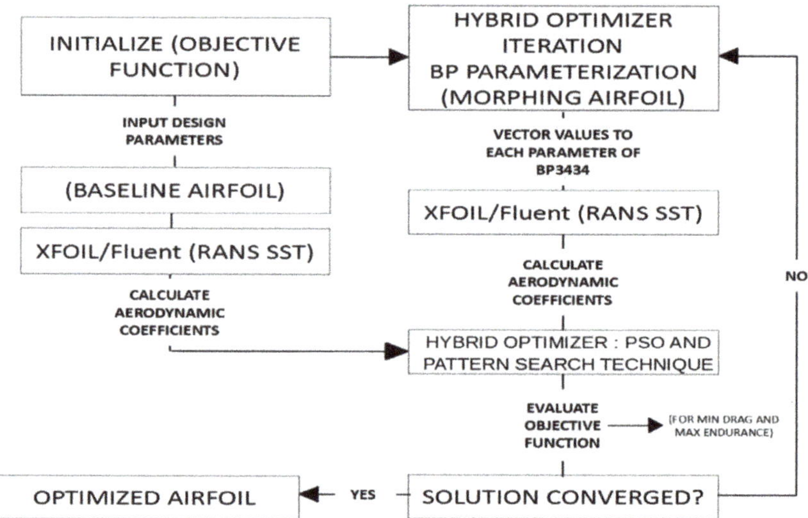

Figure 2. Schematics of the optimization procedure.

2.1. Objective Function Definition

Numerous researchers have stated that there is always a trade-off between aerodynamic coefficients and performance parameters in aerodynamic design problems [66]. A difficult balance exists between increasing aircraft performance and meeting design constraints efficiently in practical aerodynamic design problems. For example, the objective of aerodynamic optimization is lift maximization, drag minimization or maximization of the lift-to-drag ratio while keeping C_L and/or C_D constant as constraints. Likewise, various airfoil shapes are optimized for their morphing by considering an increased lift-to-drag ratio and, at the same time, for meeting the given constraints. Other airfoil shape optimizations may involve those of multiple aerodynamic coefficients based on performance parameters such as range, loiter and control. Such optimization techniques can be applied to solve complex aerodynamic design problems for a broad range of flight conditions.

In addition, the design problem shown in this paper must consider the extended range of flight operating conditions that a UAV is expected to encounter within its flight envelope. As mentioned above, this study considers the optimization of the airfoil shape for drag minimization and for endurance maximization as individual target objective functions.

The mathematical formulation of the drag minimization objective function is the following.

$$minimize\ C_D(x);\ x\ \epsilon\ (airfoil\ set) \tag{1}$$

Subject to

$$C_{L_{max}}\ (x,\ \alpha,\ M\ \geq)1{\cdot}608C_{L_{morph}} \geq C_{L_{min,\ baseline}}(x,\ \alpha,\ M)$$

where

$$x \in (airfoil\ set, the\ vector\ of\ airfoil\ shape\ function\ design\ variables)$$

The lift coefficients of the UAS-S45 are calculated for each angle of attack using design parameters mentioned in a previous paper of our team [67]. The constraints implemented are both maximum lift coefficient and lift coefficient of baseline airfoil at each angle of attack. The optimization runs are executed with a minimum lift coefficient at each angle of attack. Constraints are applied by penalty functions to the objective function to ensure optimum performance is obtained at each given angle of attack.

For the design of a long-endurance UAV, Equations (2) and (3), which are needed to determine the endurance (E) of a propeller-powered UAV, are well established using the Breguet formula [68].

$$E = \frac{\eta_{pr}}{c_p} \frac{C_L^{\frac{3}{2}}}{C_D} \sqrt{2\rho S} \left(\frac{1}{\sqrt{W_f}} - \frac{1}{\sqrt{W_i}} \right) \tag{2}$$

where c_p is the specific fuel consumption, η_{pr} is the propeller efficiency, ρ is the free stream density, S is the wing planform area, W_i is the initial weight of the aircraft, W_f is the final weight of the aircraft and C_L and C_D are the lift and drag coefficients, respectively.

The parameters that do not affect the airfoil shape design in the endurance performance can be neglected in Equation (2), and the parameters that can affect the airfoil shape design are only the two aerodynamic coefficients C_L and C_D; therefore, Equation (2) can be reformulated as

$$E_a = \frac{C_L^{\frac{3}{2}}}{C_D} \tag{3}$$

Therefore, an optimized airfoil maximizes the lift and minimizes the drag coefficients in the above Equation (3); therefore, the overall endurance aerodynamic efficiency is maximized. The weighted average of the flight envelope is used to find out the time that the UAV would have to spend in different flight regimes. Since the present study concerns the surveillance UAS-S45 optimization, it is assumed that 80% of total flight time is spent in loiter flight conditions, which require maximum endurance. Therefore, the objective function to maximize the endurance aerodynamic efficiency with the same constrained optimization is given by the following equation:

$$\text{maximize } E_a(x) = \frac{C_L^{\frac{3}{2}}}{C_D} \tag{4}$$

where
$$x \in (airfoil\ set, the\ vector\ of\ airfoil\ shape\ function\ design\ variables)$$

2.2. Parameterization Strategy

Parameterization of the airfoil shapes for the UAS design and optimization requires a mathematical formulation, and it becomes an important part of the overall optimization process. Studies have shown that the choice of shape parameterization technique has a strong influence on the solution accuracy, robustness and computational time of the overall optimization process [69]. In order to obtain an optimal aerodynamic solution, the shape function must be directly related to the airfoil geometry, have a flexible design space, and be robust for the control of all parameters. Some of the well-known methods are the Discrete Based Approach, the Bezier curves, the B-Spline or NURBS curves, the Cubic Spline and the Free-Form Representation [69,70]. The drawbacks of these methods are that they do not use airfoil shape parameters, they require a large number of design variables and they often provide inaccurate shapes for the leading edge and trailing edge of an airfoil.

These limitations have been partly removed by using the polynomial-based functions, such as PARSEC and CST. "PARSEC" is characterized by eleven design coefficients that control airfoil shape parameters, and they are the: (a) edge radius, (b) upper crest abscissa, (c) upper crest ordinate, (d) upper crest curvature, (e) lower crest abscissa, (f) lower crest ordinate, (g) lower crest curvature, (h) trailing edge ordinate, (i) trailing edge thickness, (j) trailing edge direction and (k) trailing edge wedge angle. Several studies have shown PARSEC method's superior performance with respect to the Discrete Based Approach, based on their convergence rate, flexibility and epistasis, which is due to the nonlinear dependency of the objective function on the design parameters [71]. However, it has also been found that the leading edge and trailing edge design coefficients in this parameterization method do not provide accurate airfoil shape; therefore, implementing morphing design using only these PARSEC design coefficients might further increase the airfoil shape complexity.

Therefore, a combined approach, the Bezier-PARSEC (BP) parameterization method, was developed [72] that included the advantages of Bezier with PARSEC combination, and it was found to be suitable for this study. The BP parameterization was implemented in this paper with the aim to increase the solution accuracy, flexibility and efficiency of morphing leading edge and trailing edge optimizations. More significantly, this technique provides the large search space needed for morphing leading edge and trailing edge design, and it reduces the computational time.

The BP method has been further divided into two parameterization sub-methods: BP3333 and BP3434. Four third-order Bezier curves are used to represent the airfoil shape in the BP3333 parameterization sub-method. The BP3434 parameterization has a 3rd-degree edge thickness curve, a 4th-degree trailing edge thickness curve, a 3rd-degree leading edge camber curve and a 4th-degree trailing edge camber curve. Due to the smaller number of degrees of freedom in the BP3333 method with respect to those of the BP3434 method, specifically at the trailing edge and at the leading edge, the BP3434 method was chosen in this research.

A fourth-degree Bezier curve is given by:

$$x(u) = x_0(1-u)^2 + 4x_1 u(1-u)^3 + 6x_2 u^2(1-u)^2 + 4x_3 u^3(1-u) + x_4 u^4 \tag{5}$$

and

$$y(u) = y_0(1-u)^2 + 4y_1 u(1-u)^3 + 6y_2 u^2(1-u)^2 + 4y_3 u^3(1-u) + y_4 u^4 \tag{6}$$

The parameterization is controlled by 15 parameters: 10 aerodynamic and 5 Bezier parameters. Figure 3 shows the graphical representations of these parameters.

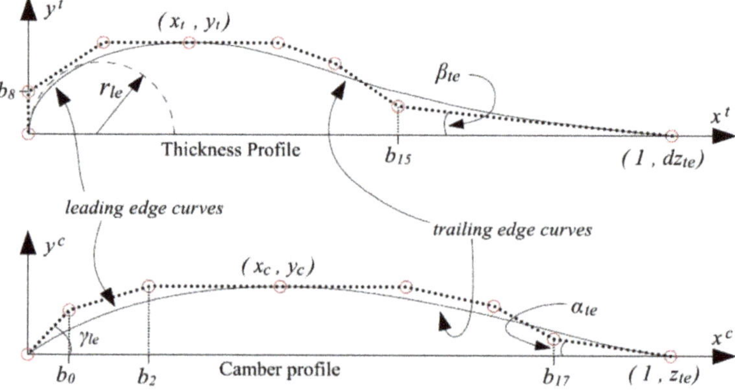

Figure 3. BP3434 method parameter definition. The BP3434 parameterization has a 3rd-degree edge thickness curve, a 4th-degree trailing edge thickness curve, a 3rd-degree leading edge camber curve and a 4th-degree trailing edge camber curve.

The control points definitions for the leading and trailing edge Bezier curves are given in Table 2 for the thickness profile curve, and in Table 3 for the camber profile curve [73]. The upper and lower bound values of the parameters used in the BP parameterization optimization method calculated in this study for the UAS-S45 are presented in Table 4.

Table 2. Thickness profile curve control points definition.

Leading Edge		Trailing Edge	
$x_0 = 0$	$y_0 = 0$	$x_0 = x_t$	$y_0 = y_t$
$x_1 = 0$	$b_1 = b_8$	$x_1 = \left(7x_t + 9b_8^2/2r_{le}\right)/4$	$y_1 = y_t$
$x_2 = \frac{-3b_8^2}{2r_{le}}$	$y_2 = y_t$	$x_2 = 3x_t + 15b_8^2/4r_{le}$	$y_2 = (y_t + b_8)/2$
$x_3 = x_t$	$y_3 = y_t$	$x_3 = b_{15}$	$y_3 = dz_{te} + (1 - b_{15})\tan(\beta_{te})$
		$x_4 = 1$	$y_4 = dz_{te}$

Table 3. Camber profile curve control points definition.

Leading Edge		Trailing Edge	
$x_0 = 0$	$y_0 = 0$	$x_0 = x_c$	$y_0 = y_c$
$x_1 = b_0$	$y_1 = b_0\tan(\gamma_{te})$	$x_1 = \frac{1}{2}(3x_c - y_c cot(\gamma_{le}))$	$y_1 = y_c$
$x_2 = b_2$	$y_2 = y_c$	$x_2 = \frac{1}{6}(-8y_c cot(\gamma_{le}) + 13x_c)$	$y_2 = \frac{5}{6}y_c$
$x_3 = x_c$	$y_3 = y_c$	$x_3 = b_{17}$	$y_3 = z_{te} + (1 - b_{17})\tan(\alpha_{te})$
		$x_4 = 1$	$y_4 = z_{te}$

Table 4. Values of the upper and lower bounds of parameters used in the BP method.

Parameter	r_{le}	β_{te}	x_t	y_t	dz_{te}	b_8	b_{15}	α_{te}	z_{te}	γ_{le}	x_c	y_c	b_0	b_2	b_{17}
Lower	-0.036	0.0009	0.0009	0	0.045	0.18	0	0.135	0.045	0	0.009	0.09	0	0	0
Upper	-0.003	0.3	0.9	0.03	0.6	1.5	0.6	1.2	0.45	0.003	1.5	1.8	2.1	2.7	2.7

2.3. Description of Flow Solvers

Various solvers have been employed in aerodynamic optimization studies, such as the Fully Potential Flow, Coupled Boundary-Layer, Euler and Viscous Navier–Stokes solvers [74]. The type of solver chosen by a user depends upon the type of optimization problem. Results obtained with the flow solver must be consistently accurate in order for the optimization to be considered highly adequate.

For this aerodynamic research, incompressible flow solvers are needed as they can perform fast computation, and also can provide high accuracy results. Therefore, in this paper, two types of aerodynamic solvers are used for the aerodynamic optimization: a Panel-based method and a Reynolds-Averaged Navier–Stokes equations (RANS) solver. The choice to keep the computational cost low for an airfoil analysis leads directly to the use of the XFOIL code, which is known for its very good combination of execution speed and accuracy of results. Furthermore, in order to obtain higher accuracy results for solving the viscous boundary layer and the flow separation than the XFoil solver, a well-established high-fidelity Computational Fluid Dynamics (CFD) solver, Ansys Fluent was also used in this morphing airfoil analysis.

2.3.1. XFoil

XFoil is a code for airfoil design and analysis consisting of inviscid, inverse and viscous formulations [75]. The viscous formulation can be used to calculate free or imposed forced flow transition to handle transitional separation bubbles, to calculate aerodynamic coefficients and to cope with moderately trailing edge separation. It uses an approximate e^N envelope method to calculate the flow transition. The turbulence level settings are kept as default with free transition features. Both the boundary layer and the wake parameters are calculated with a two-equation integral boundary layer formulation as shown in Equations (8) and (9), respectively [76].

$$\frac{d\theta}{d\zeta} + \left(2 + H - M_e^2\right)\frac{\theta}{u_e}\frac{du_e}{d\zeta} = \frac{C_f}{2} + \left\{\frac{v_0}{u_e}\right\} \tag{7}$$

$$\theta\frac{dH^*}{d\zeta} + (2H^{**} + H^*(1-H))\frac{\theta}{u_e}\frac{du_e}{d\zeta} = 2C_D - H^*\frac{C_f}{2} + \left\{(1-H^*)\frac{v_0}{u_e}\right\} \tag{8}$$

where shape factors H^* and H^{**} are defined in the following equations

$$xH^* = \frac{\int_0^\infty \frac{u}{U}\left(1 - \left(\frac{u}{U}\right)^2\right)dy}{\int_0^\infty \frac{u}{U}\left(1 - \left(\frac{u}{U}\right)\right)dy} \tag{9}$$

and

$$H^{**} = \left(\frac{0.064}{H_k - 0.8} + 0.251\right)M_e^2 \text{ with } H_k = \frac{H - 0.290M_e^2}{1 + 0.113M_e^2}$$

At high angles of attack where stall occurs, XFoil has difficulties in giving a converging solution for the airfoil analysis. To obtain the convergence of the solution, non-converged solutions are given a high penalty by setting the fitness function arbitrarily large via a penalty function, and thus they are eventually eliminated during the design process. The solver tolerance value used for this penalty function calculation is 0.002.

Many researchers have employed XFoil software in morphing airfoil studies. In one of these studies, the results obtained using XFoil were compared with those obtained using the Reynolds-Averaged Navier–Stokes (RANS) solvers for camber morphing, and only a slight difference between the results was observed [77].

2.3.2. CFD Fluent Solver

The validation of solvers and mesh settings is needed before carrying out the optimization studies. The validation studies were performed at a Reynolds number of 2.4×10^6 and a Mach number of 0.10. The range of angles of attack was considered from $0°$ to $16°$ to include the stall angle. The aerodynamic coefficients and the pressure distribution were obtained and further used for solution validation. A C-shaped computational domain with structured grids, as seen in Figure 4, was used in domain discretization after the performance of the mesh convergence test. The turbulence model used in this study was the Transition $(\gamma - Re_\theta)$ SST. This model uses a combination of SST K-ω coupled with intermittency γ and transition onset Reynolds number. Re_θ is the critical Reynolds number where the intermittency starts [78]. Four transport equations of the Transition $(\gamma - Re_\theta)$ SST model are given below:

$$\frac{\partial(\rho k)}{\partial t} + \frac{\partial(\rho U_j k)}{\partial x_j} = P_k - D_k + \frac{\partial}{\partial x_j}\left[(\mu + \sigma_k\mu_t)\frac{\partial k}{\partial x_j}\right] \tag{10}$$

$$\frac{\partial(\rho\omega)}{\partial t} + \frac{\partial(\rho U_j\omega)}{\partial x_j} = P_\omega - D_\omega + \frac{\partial}{\partial x_j}\left[(\mu + \sigma_\omega\mu_t)\frac{\partial\omega}{\partial x_j}\right] + 2(1 - F_1)\frac{\rho\sigma_{\omega 2}}{\omega}\frac{\partial k}{\partial x_j}\frac{\partial\omega}{\partial x_j} \tag{11}$$

$$\frac{\partial(\rho\gamma)}{\partial t} + \frac{\partial(\rho U_j\gamma)}{\partial x_j} = P_\gamma - E_\gamma + \frac{\partial}{\partial x_j}\left[\left(\mu + \frac{\mu_t}{\sigma_\gamma}\right)\frac{\partial\gamma}{\partial x_j}\right] \tag{12}$$

$$\frac{\partial(\rho\overline{Re_{\theta t}})}{\partial t} + \frac{\partial(\rho U_j\overline{Re_{\theta t}})}{\partial x_j} = P_{\theta t} + \frac{\partial}{\partial x_j}\left[\sigma_{\theta t}(\mu + \mu_t)\frac{\partial\overline{Re_{\theta t}}}{\partial x_j}\right] \tag{13}$$

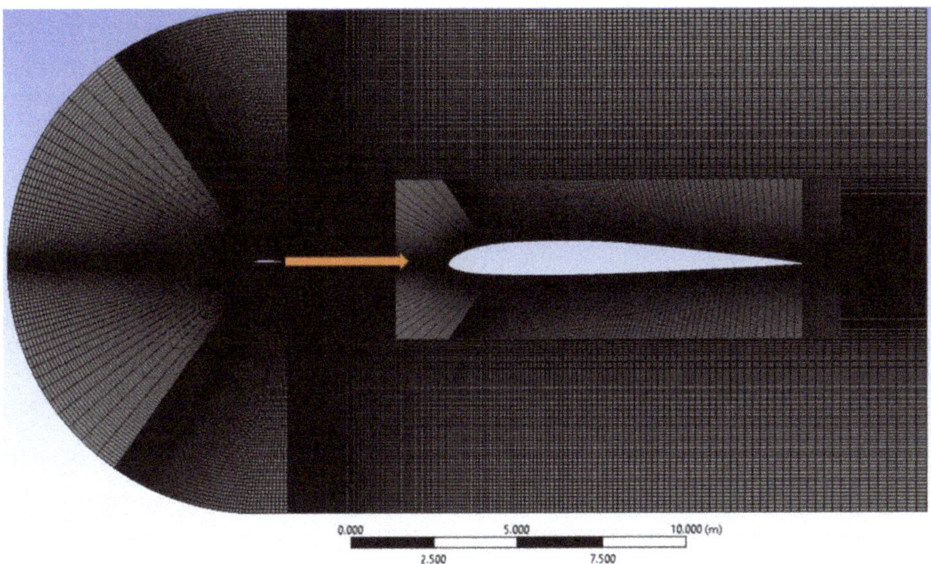

Figure 4. Grid around the airfoil used in Reynolds-averaged Navier–Stokes (RANS) simulations.

The spatial discretization used in the solution is of the second order for all the turbulence statistical parameters, including the momentum, the turbulent kinetic energy, the specific dissipation rate, the intermittency and the Reynolds momentum thickness.

ANSYS Fluent was used in this study to carry out the numerical analysis. The meshing was done using ANSYS Mesher, as shown in Figure 4. The rectangular computational domain was designed around the airfoil geometries. The outlet length was selected to be 30c, which was adequate to allow for the full development of the wake flow, based on the size of the computational domain adopted in previous studies. The distance from the domain inlet to the airfoil was 10c to prevent the inlet boundary from unphysically impacting the induction field upstream of the airfoil. The high-quality grids were generated with dense grids in the boundary layer of the airfoil with a gradual decrease of grid cells away from the airfoil surface. The first layer's thickness was calculated based on inlet velocity and adopted accordingly to capture the laminar and transitional boundary layer regions. Furthermore, the grid sensitivity analysis was performed according to the number of cells in the computational domain.

A freestream velocity of 34 m/s with a turbulence rate of 0.01 percent was imposed at the inlet. The surface of the airfoil was modelled as a zero-roughness no-slip wall, and its values were executed on the grid domain, which prescribed its motion relative to the rest of the computational domain. A symmetry boundary condition was applied to achieve a parallel flow at the top and bottom of the domain by assuming zero normal velocity and zero normal gradients of the flow quantities. A zero static gauge pressure was applied at the exit of the domain. Iterations were finalized when all scaled residuals were below 1×10^6.

2.4. Optimization Algorithm

An intelligent search algorithm is essential for the direct numerical optimization of airfoil design. This algorithm operates iteratively and utilizes the inputs of the shape parameterization method to define the airfoil shape, and then it uses the flow solver to calculate the aerodynamic coefficients. Its efficiency is assessed according to its ability to provide a global optimal solution with reasonable computational resources. Therefore, the choice of such an algorithm influences the solution convergence and its feasibility.

Researchers have implemented various algorithms, such as the Gradient-Based, Adjoint-Based and Evolutionary algorithms with the aim to investigate and solve various optimization problems [74]. These algorithms have been considered to solve different aerodynamic optimization problems based on their advantages and disadvantages. The hybrid optimizer based on the Particle Swarm Optimization (PSO) algorithm was used in this paper for the UAS-S45 optimization, and it was further combined with the Pattern Search algorithm in order to enhance the solution convergence and its refinement.

Evolutionary algorithms have attracted the attention of shape designers due to their characteristics, such as their suitability for modelling discontinuous shape functions, obtaining global optimal solutions, ease of parallel computing, etc. During morphing airfoil optimization, gradient-based algorithms converge fast, but they only cover a small area out of a large airfoil search space, and thus a local minimum solution instead of global minimum solution is found. It is also difficult to find the gradient of the non-linear flow fields for gradient-based optimizers.

Therefore, a PSO algorithm is used, based on a simplified social behavior closely related to the swarming theory, where the solutions are represented by a set of particles that heuristically navigate through a design space [79]. The efficiency of PSO algorithms over genetic algorithms is due to their independence from parameters, such as crossovers and mutations; instead, the solution is updated by sharing its information amongst its populations of particles.

2.4.1. PSO Algorithm

In the PSO algorithm, each particle is a solution to a given optimization problem and is composed of two vectors: "position" and "velocity". A position vector x_i^n is used to store the positioning of the particles in the given dimensional space. The velocity vector v_i^n is updated using the following equation for each iteration k:

$$v_i^n(k) = [v_i^n(k-1) + c_1 r_1 (pbest_i^n - x_i^n(k-1)) + c_2 r_2 (pbestg_i^n - x_i^n(k-1))] \qquad (14)$$

where x_i^n = position vector; v_i^n = randomly generated velocity vector of the particles at search initialization; $pbest_i^n$ = vector representing the best solution achieved by the particle; $pbestg_i^n$ = vector representing the "global best solution" collectively achieved by the swarm; c_1 and c_2 = stochastic acceleration terms that pull each particle toward $pbest_i^n$ and $pbestg$ positions respectively; r_1 and r_2 = random numbers in the uniform range [0, 1];

The update of the position vector at the kth iteration is given by the following equation:

$$x_i^n(k) = x_i^n(k-1) + v_i^n(k) \qquad (15)$$

The PSO procedure can be outlined as follows:

1. Generate an initial swarm with arbitrary values of the particle position and arbitrary initial velocities in the n-dimensional search space;
2. Evaluate the fitness of each particle of the swarm by searching a new velocity vector applied to each individual particle; the new velocity vector is influenced by its best position, the swarm's best position and its previous velocity;
3. Update the velocity of the particles as shown in Equation (14) and the position of the particles as shown in Equation (15) and evaluate the new fitness function;
4. Repeat steps 2 and 3 until the maximum number of iterations is reached or until the results no longer improve for a given number of iterations.

Pattern Search (PS) is a direct search optimization method that is used to minimize a function by comparing its value at each iteration with its value at its previous iteration in a finite set of trial points. The method is outlined as follows:

1. Select the initial solution;
2. Explore the direction search by evaluating each objective function value, one at a time;

3. Evaluate the objective function. If the objective function is minimized, then update it; otherwise, do not update it;
4. Evaluate the next values of the objective function following the previous steps 2 to 3;
5. After exploring the search of all possible values of the objective function and the minimization of the overall objective function, the PS method follows a pattern move; and
6. If the overall objective function has not been minimized, repeat the process.

2.4.2. Hybrid Optimization Scheme

A hybrid optimizer is used in this study, as shown in Figure 5. In this figure, the hybrid optimization implies that the PSO-based optimization process is applied firstly to its variables to obtain their optimized values. Since the first optimization process may not be accurate as it would result in low variations of values to be optimized, a second optimization process would then be applied to the results of the first optimization process. In this case, PSO, a non-elitist algorithm, is then followed by Pattern Search, using hybrid optimization. The PSO and Pattern Search can then be used to find a local optimum in a limited search area; the use of both PSO and PS algorithms through hybrid optimization ensures better results than a single optimization process. The independent variables given as inputs to the optimization process are also limited due to their upper and lower bounds. These bounds ensure that the search space for both the PSO and Pattern Search algorithms remains within them. During the PSO algorithm execution, if a solution is found outside the bounds of the search space, any such solution outside these bounds will have its value replaced by that of its given upper and/or lower bounds, without any influence on other solutions. The hybrid optimizer's procedure is shown in Figure 5.

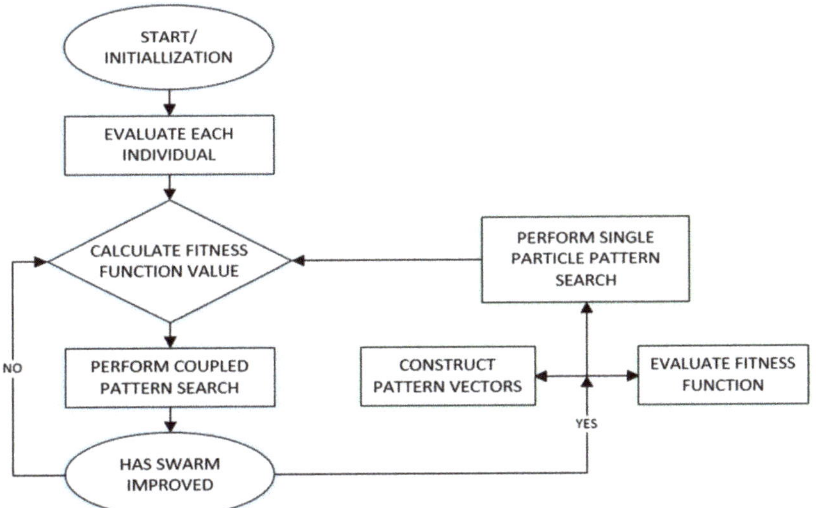

Figure 5. Schematics of the Hybrid Optimizer/ Particle Swarm Optimization (PSO) algorithm combined with the Pattern Search algorithm.

3. Discussion and Results

The aerodynamic optimization considered in this study applies to both Droop Nose Leading Edge Morphing (DNLE) and Morphing Trailing Edge (MTE) configurations. To utilize the full potential of the "morphing" concept, a baseline shape optimization was carried out followed by its morphing shape design at various flight conditions given in Table 5. These cases are based on various altitudes and Reynolds numbers that are computed for the airspeed of 34 m/s. These given flight conditions are chosen because of the fact that the UAS-S45 can reach altitudes of 20,000 ft and a stall speed of 34 m/s. The design framework uses endurance maximization as an optimization function to enhance the performance of the UAS-S45 over a large part of its flight regimes. Figure 6 shows the flight envelope of UAS-S45. From a physical standpoint, every flight condition is defined by a specific Mach number, altitude (Reynolds number and temperature are calculated according to ISO standard atmosphere) and a fixed lift coefficient. The present study considered the optimization of the leading edge and trailing edge of the UAS S45 conventional wing that could be both morphed. The difference of performances obtained between the baseline airfoil and the morphed one demonstrate the overall benefits of morphing airfoil design. Figure 7 shows the design of Droop Nose Leading Edge (DNLE) and Morphing Trailing Edge (MTE) airfoil configurations versus the baseline airfoil. Figure 7a shows the basic concept of the morphing wing with combined morphing leading and trailing edge deflections and Figure 7b shows two cases of MTE designs.

Table 5. Operating conditions for design problem of UAS-S45 at the airspeed of 34 m/s.

Flight Condition	Altitude (Feet)	Reynolds Number
Flight Condition I	0	2.40×10^6
Flight Condition II	10,000	1.88×10^6
Flight Condition III	20,000	1.45×10^6

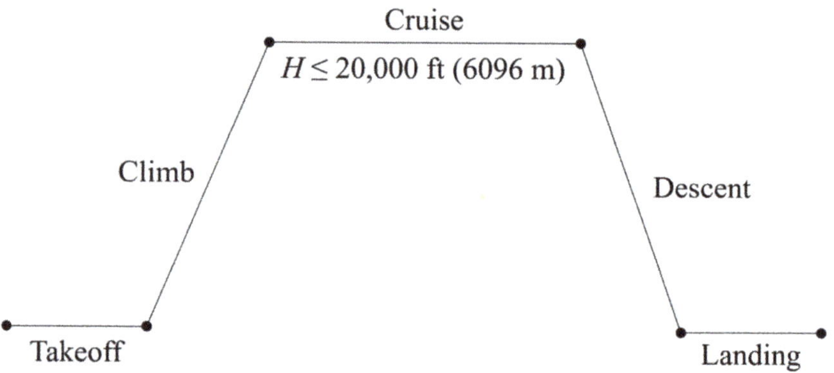

Figure 6. Schematic of the flight regimes of UAS-S45.

Figure 8 shows the aerodynamic coefficients C_L and C_D variations of the baseline UAS-S45 root airfoil for two Reynolds numbers Re = 1.2×10^6 and Re = 2.4×10^6, which were calculated using the XFoil and the Transition ($\gamma - Re_\theta$) SST in Ansys/Fluent software. As seen in Figure 8, the lift coefficients' variations with the angle of attack were predicted accurately, while small differences were found in drag coefficient variations with the angle of attack. These differences in the drag coefficients' variations are found largely at angles of attack higher than 4° because the pressure-based drag coefficient becomes the dominant component of the total drag coefficient and flow separation conditions result in a decreased friction-based drag coefficient. This result implies that drag forces are very sensitive to turbulence effects and the Transition ($\gamma - Re_\theta$) SST model can solve the turbulence problems more accurately than the XFoil solver.

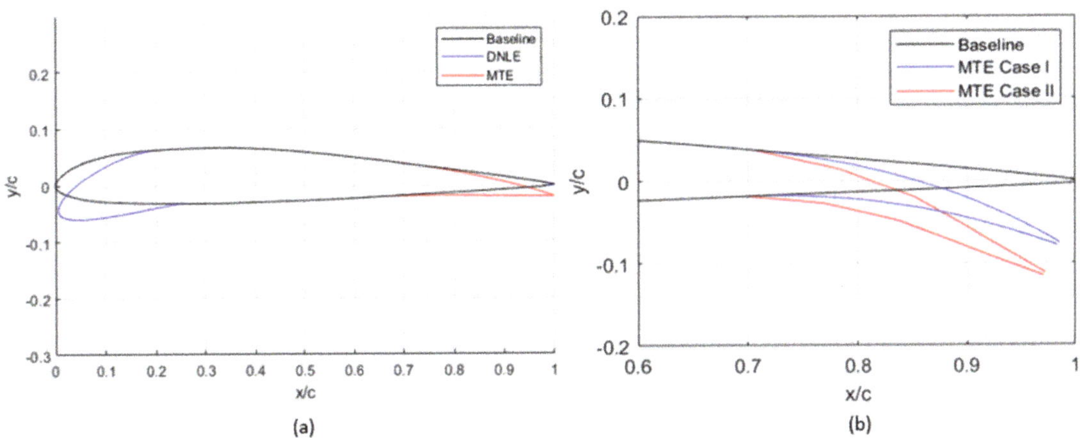

Figure 7. Design of Droop Nose Leading Edge (DNLE) and Morphing Trailing Edge (MTE) airfoil configurations versus the baseline airfoil. (**a**) the basic concept of the morphing wing with combined morphing leading and trailing edge deflections; (**b**)the two cases of MTE designs.

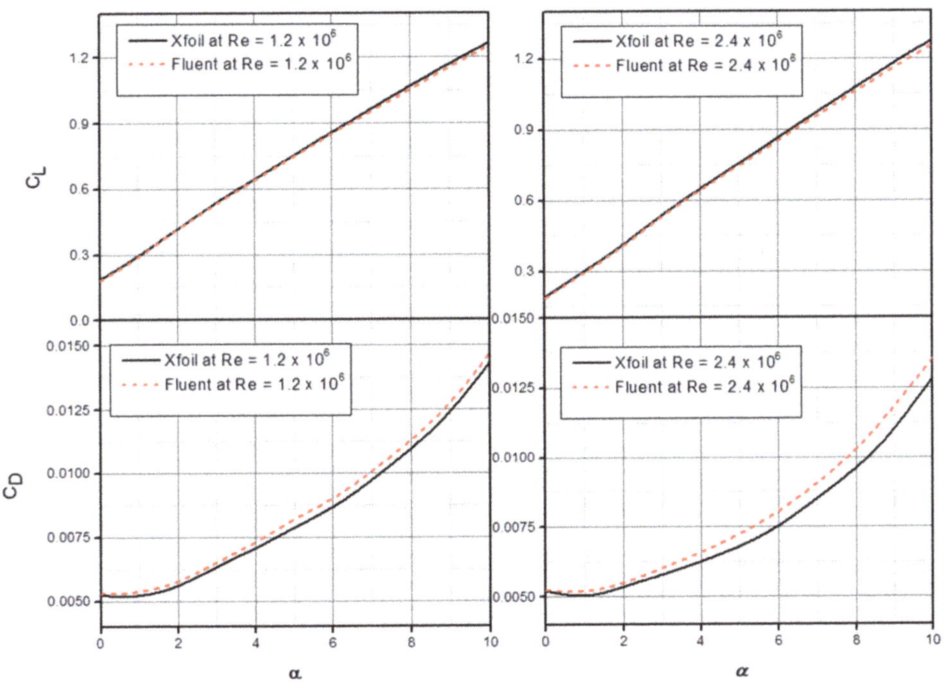

Figure 8. Aerodynamic lift and drag coefficient variations using the XFoil and Fluent solvers.

Figure 9a shows the friction coefficient variations with the chord for the angles of attack of $0°$, $2°$ and $6°$ obtained by use of the Transition $(\gamma - Re_\theta)$ SST model; this figure indicates very good accuracy in predicting the flow separation phenomena as the skin friction coefficient increases slightly with the increase in angle of attack. Figure 9a,b also shows the initial flow separation, the separated region and its reattachment. It is seen in

Figure 9a that the laminar-to-turbulent transition region is located at 18% of the chord (0.18c) at an angle of attack of 6° and at 43% of the chord (0.43c) at an angle of attack of 2°. The variations of the skin friction coefficient along the chord were assessed at different angles of attack using the Transition $(\gamma - Re_\theta)$ SST model in Figure 9.

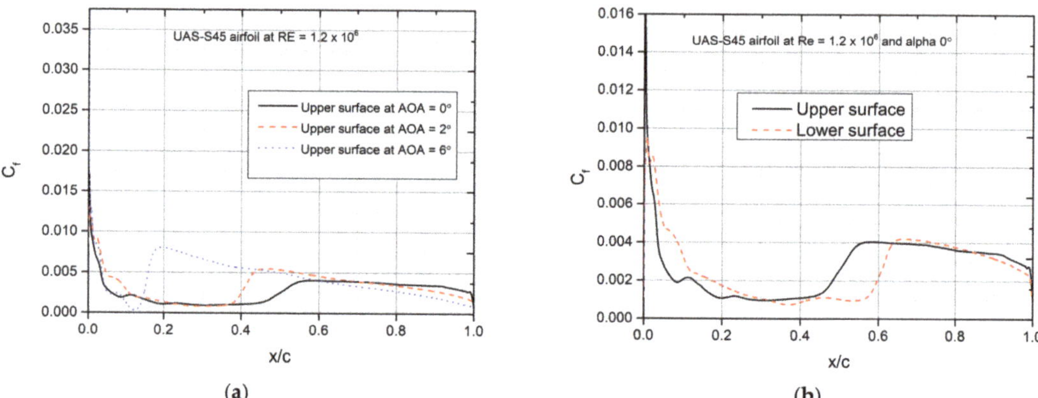

Figure 9. Skin friction coefficients variations with the chord obtained at different angles of attack. (**a**) the friction coefficient variations with the chord for the angles of attack of 0°, 2° and 6°; (**b**) the initial flow separation, the separated region and its reattachment for upper and lower surface.

The velocity magnitude contours plots at three angles of attack (4°, 10°, 16°) at Re = 2.4×10^6 are shown in Figure 10. The turbulent intensities play an important role in understanding the flow behavior over an airfoil, and they are shown in Figure 10 for the angles of attack of 4° and 10°.

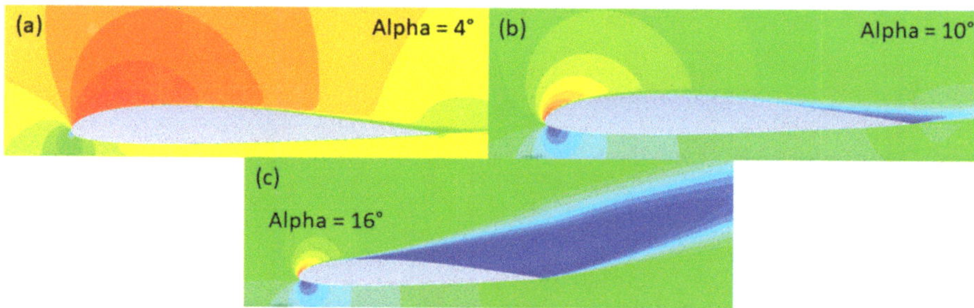

Figure 10. Velocity magnitude contour plots at three angles of attack at Re = 2.4×10^6. (**a**) the contour lines at low angles of attack such as 4°; (**b**) the contour lines at low angles of attack such as 10°; (**c**) a fully separated flow at the angle of attack of 16°.

The contour lines remain attached to the airfoil at low angles of attack such as 4°, as shown in Figure 10a; the flow starts to separate at the trailing edge of the airfoil at an angle of attack of 10° as shown in Figure 10b and changes into a fully separated flow at the angle of attack of 16° as shown in Figure 10c. It is important to understand that boundary layer separation takes place at an angle of attack of approximately 10°. The separation region shown in Figure 10c causes increased drag as it induces a large wake that completely changes the flow downstream of the point of separation. It is clear that the turbulent intensity is minimal at the airfoil leading edge at a low angle of attack of 4°, as seen in Figure 11a, while at the angle of attack of 10°, the turbulent intensity increases over the airfoil surface, including at its leading edge.

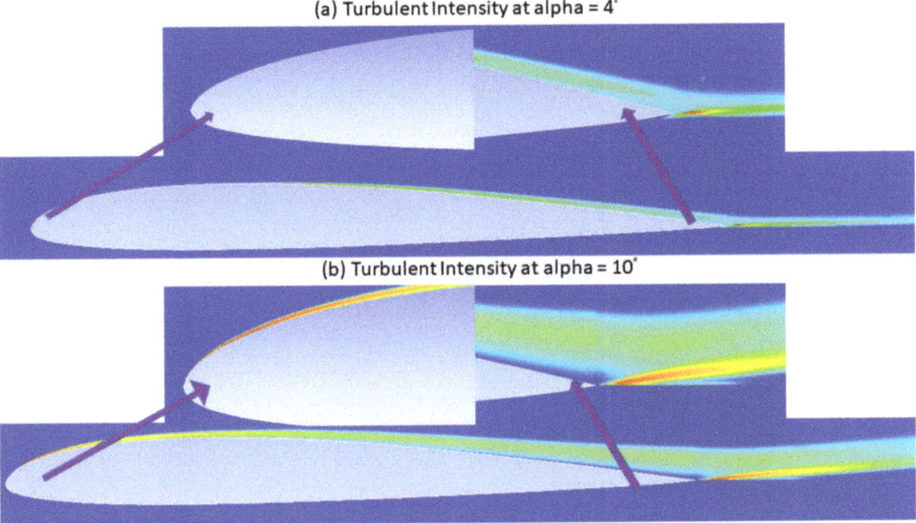

Figure 11. Contour plots of turbulent intensity comparison for baseline airfoil at the airspeed of 34 m/s. (**a**) the turbulent intensity at a low angle of attack of 4°; (**b**) the turbulent intensity at a low angle of attack of 10°

3.1. Optimization of Morphing Leading Edge

The Droop Nose Leading Edge (DNLE) optimization was performed to satisfy the objective functions defined in Equation (1) for drag minimization and in Equation (5) for endurance maximization. The optimization processes were carried out for two cases of DNLE control point deflections, defined as "Opt. Case I" and "Opt. Case II".

In "Opt. Case I", a morphing DNLE is a smooth single structural layout with a morphing leading edge starting location fixed at 22% of the chord. The trailing-edge is not allowed to move, while the leading edge points are only allowed to move in the vertical direction to effectively adapt the chord-length variation of the baseline airfoil.

In "Opt. Case II", the morphing DNLE is free to change its shape anywhere from 0 to 22% of the chord. The starting point and the end point of the leading edge are used to represent the morphing DNLE shape. The parameterization of the DNLE shape was obtained by using the Bezier–PARSEC (BP) method, in which the leading edge radius of the airfoil and 5 Bezier control points were used to maintain the slope continuity within the airfoil shape.

The XFoil software was used to efficiently investigate the aerodynamics of the optimized morphing geometries. The aerodynamic performance of the DNLE was evaluated to determine its best airfoil design shape for each flight condition. The computations were performed for the three flight conditions listed in Table 5. The results were obtained in terms of aerodynamic lift and drag coefficients variations with angle of attack and pressure distributions versus the airfoil chord locations for the optimized airfoil geometries.

Figure 12a illustrates the initial (reference) versus the final (optimized or morphed) airfoil shape, while Figure 12b shows the convergence trend for the minimum drag optimization function for an optimized airfoil shape with DNLE at an angle of attack of 50. The algorithms PSO with Pattern Search converge to the optimal solution after 90 iterations. The x-axis indicates the number of function evaluations, and the y-axis shows the value of the fitness. The fitness value is the objective function value of each particle of the algorithm, expressing the performance of any airfoil shapes.

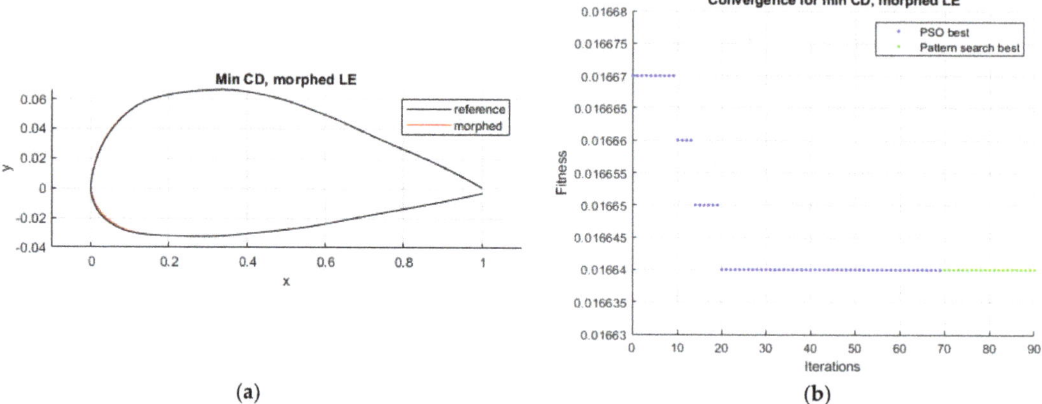

(a)　　　　(b)

Figure 12. Optimized airfoil and convergence history for drag minimization function; (**a**) the initial (reference) versus the final (optimized or morphed) airfoil shape, (**b**) the convergence trend for the minimum drag optimization function.

These results for the droop nose leading edge (DNLE) optimization are shown in Figures 13 and 14, and the UAS-S45 baseline airfoil aerodynamic coefficients were compared to those of the optimized airfoils. Figure 13a–d show the variations of lift coefficients with the angles of attack. The lift coefficient variations for all three flight conditions for Opt. Case I and Opt. Case II and the baseline airfoil in Figure 13a–c show that higher lift coefficients were obtained for Opt. Cases I and II than those of the baseline airfoil. Figure 13d shows that the DNLE designs for Opt. Case I at three different flight conditions. It is found that Opt. Case I DNLE design is the best at Flight Condition III. These optimal shapes revealed that maximum increment of the lift coefficient with respect to the baseline airfoil of up to almost 21% was found for Opt. Case I design seen in Figure 13c. In addition, an increase of 9.6% was obtained for the maximum lift coefficient, associated with a stall angle delay of 3° as seen in Figure 13c. Therefore, it was found that both Opt. Case I and Opt. Case II results are better than the baseline airfoil results. These better results can be explained by the pressure distribution change around the airfoil. The DNLE design does not cause an adverse pressure gradient as the flap configuration, but instead, it decreases the maximum speed at the airfoil leading edge, which is similar to the findings [80].

Figure 14 presents the evaluation of the lift vs. drag polar of the baseline airfoil and of the optimized airfoils for three different flight conditions. Therefore, the increased aerodynamic performance of optimized airfoils in terms of their lift versus drag variation for all three flight conditions can be seen in Figure 14a–d.

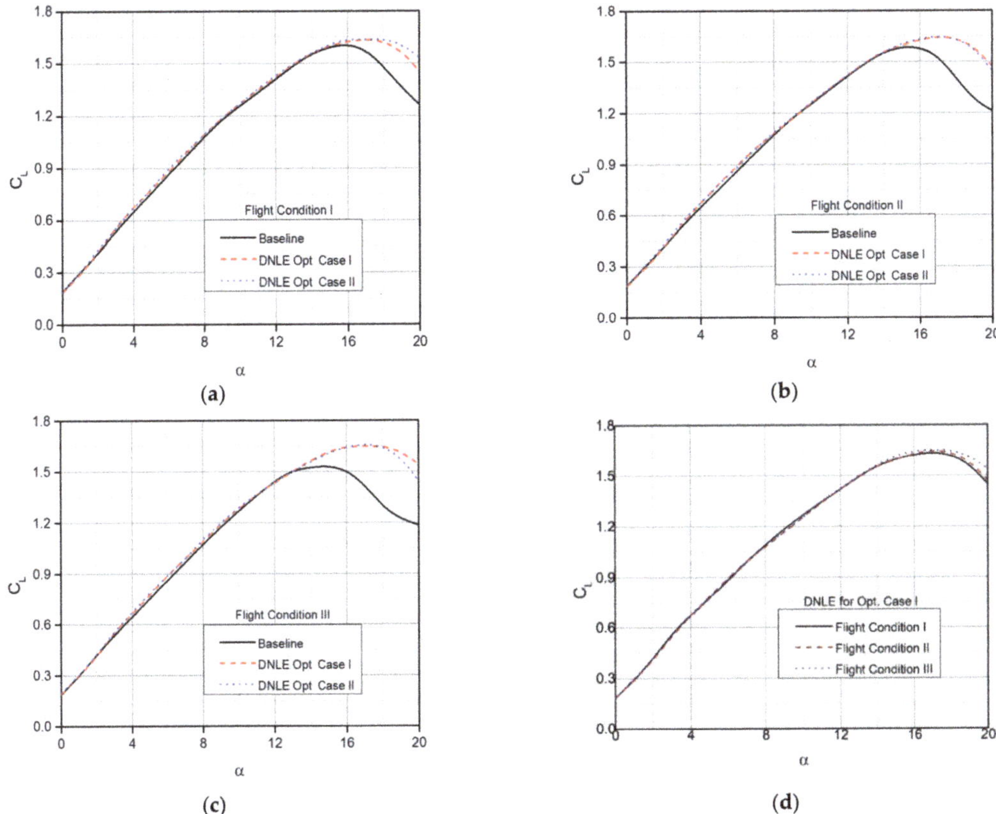

Figure 13. Lift coefficients versus angle of attack at two optimized design configurations; (**a**) Figure 13a show the variations of lift coefficients with the angles of attack at Flight Condition I (**b**) the variations of lift coefficients with the angles of attack at Flight Condition II; (**c**) the variations of lift coefficients with the angles of attack at Flight Condition III; (**d**) the lift coefficient variations for all three flight conditions for Opt. Case I.

The pressure coefficients' variations with the chord of optimized airfoils for three flight conditions are shown in Figure 15. It is clear that the major changes in pressures take place near the leading edge at its upper surface, and that the pressure peaks then remain smooth on the rest of the airfoil for all configurations in Figure 15a for an angle of attack of 4° and in Figure 15b for an angle of attack of 10°. The chord-wise pressure distribution reveals that increased performance was obtained for DNLE configurations. By comparing the chord-wise pressure distributions for different configurations at angles of attack of 10° for the baseline versus optimized airfoils, it is found that a kink in the suction peak is followed by a constant pressure distribution. The DNLE-optimized configuration at flight condition III undergoes flow separation, and reattaches shortly on the airfoil surface, as seen in Figure 15b.

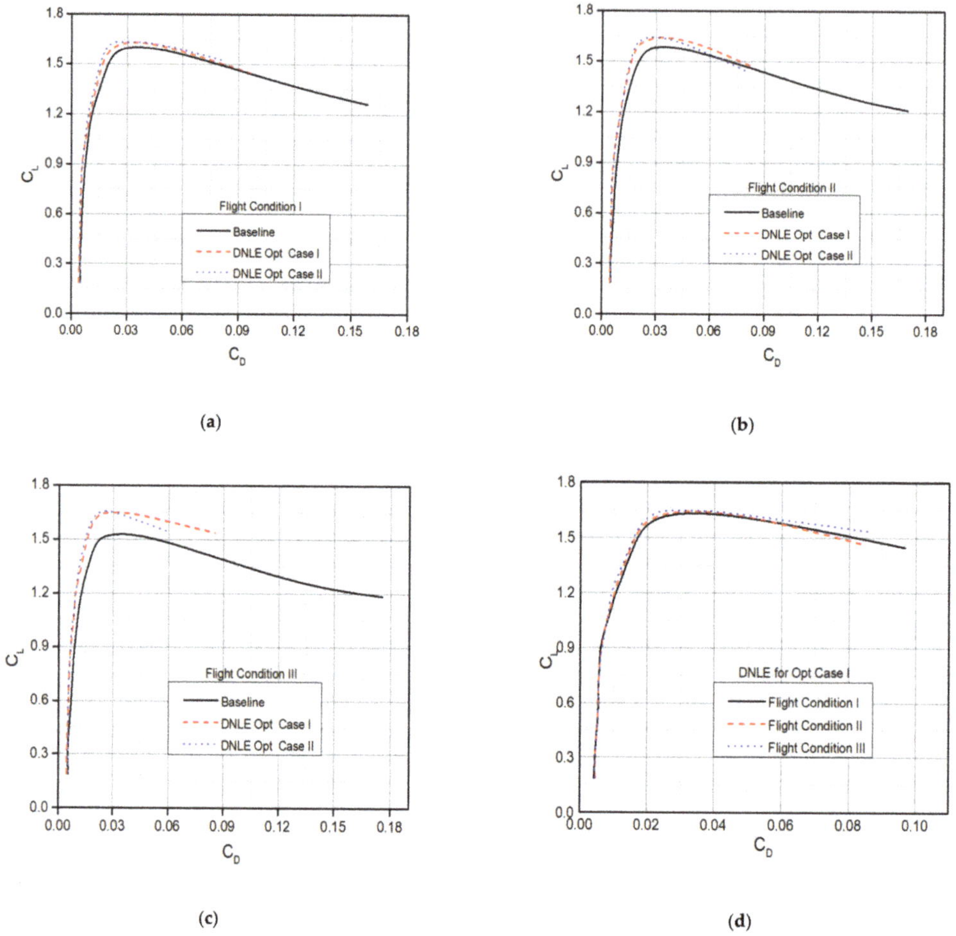

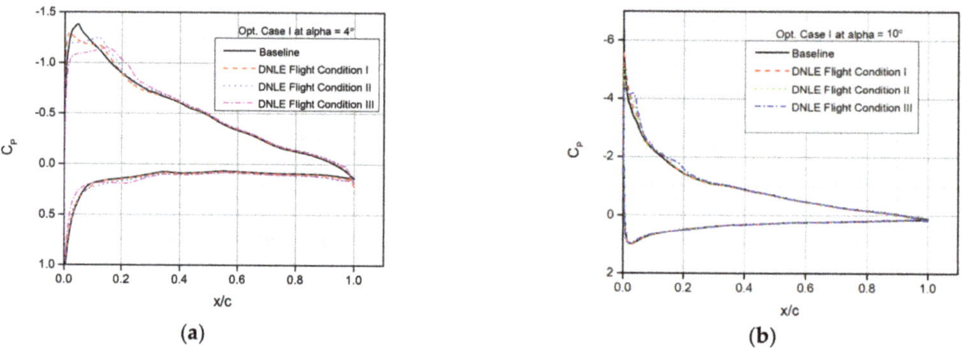

Figure 14. Aerodynamic performance for two optimized design configurations. (**a**) Figure 13a show the drag polar at Flight Condition I; (**b**) drag polar at Flight Condition II; (**c**) drag polar at Flight Condition III; (**d**) drag polar for all three flight conditions for Opt. Case I.

Figure 15. Pressure coefficients variations with the airfoil chord for Opt. Case I (**a**) at angles of attack of 4°; (**b**) at angles of attack of 10°.

Figure 16a illustrates the initial (reference) versus the final (optimized or morphed) airfoil shape, while Figure 16b presents the convergence for aerodynamic endurance optimization while utilizing the PSO algorithm with Pattern Search for the DNLE-optimized airfoil at an angle of attack of 50°. A downward trend followed by constant values is clearly visible for the function, thus indicating the approach to the global optimum.

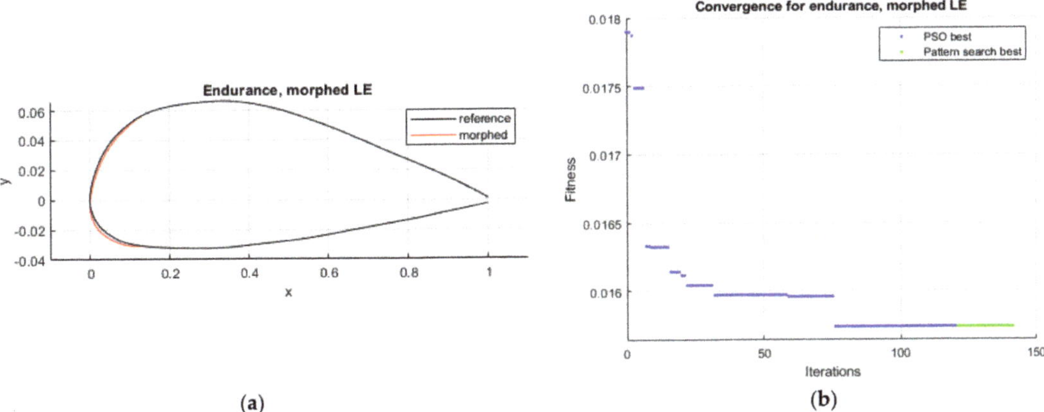

Figure 16. Optimized airfoil and convergence history for endurance maximization function. (**a**) the initial (reference) versus the final (optimized or morphed) airfoil shape; (**b**) the convergence for aerodynamic endurance optimization.

Figure 17 shows that the optimization process led to an increase in endurance maximization characterized by $C_L^{3/2}/C_D$ in the optimized DNLE airfoils for three different flight conditions. The maximum endurance capability given by $C_L^{3/2}/C_D$ for the same lift coefficient of the baseline airfoil to the lift coefficient of the morphing DNLE configuration Opt. Case I was found to be 1.17 and 1.21, respectively. Furthermore, the values of $C_L^{3/2}/C_D$ increased from 117 to 132, thus indicating better endurance performance for the UAS-S45 DNLE airfoil configurations than for its baseline airfoil at three flight conditions.

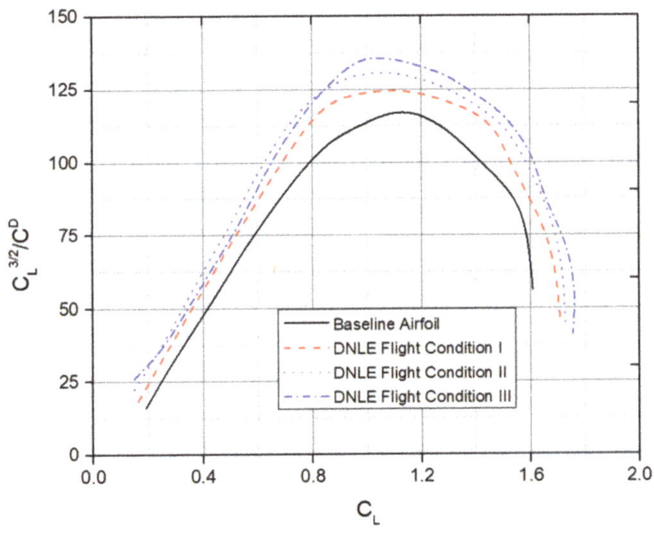

Figure 17. Comparison of the endurance for baseline and DNLE airfoils.

The flow transition behavior can be seen in the velocity contour plots of one of the DNLE airfoils, as shown in Figure 18a for an angle of attack of 4° and in Figure 18b for an angle of attack of 10°. The comparison of baseline airfoil and DNLE configurations at a 10° angle of attack shows that higher gradients are formed for the DNLE airfoil in comparison with those of the baseline airfoil, illustrated in Figure 10. However, the trailing edge flow separation region remains the same for both airfoils. In addition, the Turbulent Kinetic Energy (TKE) for the baseline airfoil and for the DNLE airfoils by use of the Transition $(\gamma - Re_\theta)$ SST turbulence model are shown in Figure 18c,d. It is evident that the leading edge is propagating the energy towards the downstream; thus a considerable decrease is observed in the wake turbulence. The TKE occurring in the baseline airfoil, which starts from the upper surface and moves towards the trailing edge was not observed in the DNLE-optimized airfoil at an angle of attack of 4°, as seen in Figure 18c.

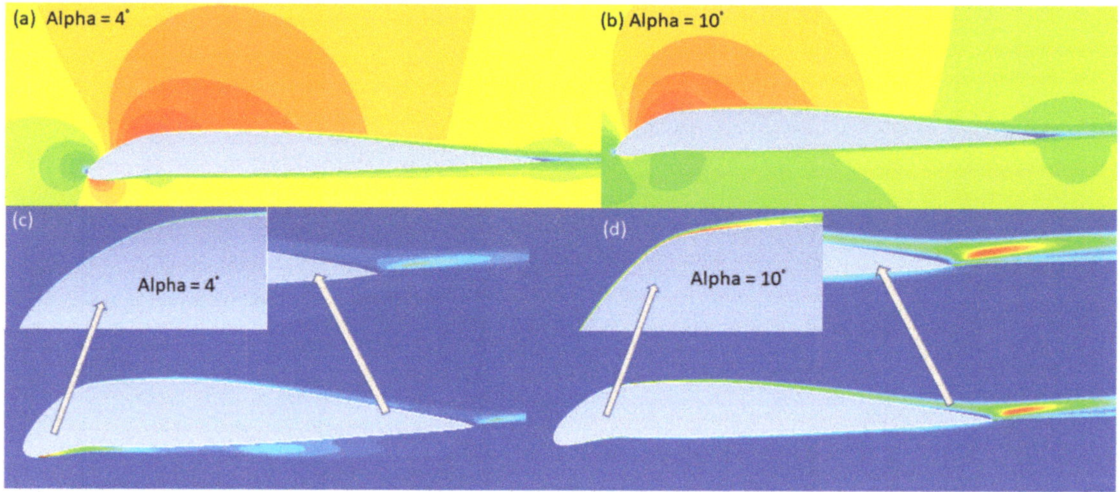

Figure 18. Velocity magnitude and turbulent kinetic energy contour plots. (**a**) Velocity contour lines at low angles of attack such as 4°; (**b**) Velocity contour lines at low angles of attack such as 10°; (**c**) the turbulent intensity at a low angle of attack of 4°; (**d**) the turbulent intensity at a low angle of attack of 10°.

3.2. Morphing Trailing Edge Optimization

The optimization process defined in Section 3.1 was implemented on the MTE airfoil by using the same two optimization functions defined by Equation (1) for drag minimization and by Equation (5) for endurance maximization. The optimization of the Morphing Trailing Edge (MTE) airfoil shape for the three flight conditions given in Table 5 was performed.

The optimization processes were carried out for two types of MTE control points. In the first type, referred to as Opt. Case I, MTE is a smooth single shape with a morphing starting point and trailing edge end point. The leading edge is not allowed to move, while the trailing edge points are only allowed to move in the vertical direction with respect to the baseline airfoil.

In the Opt. Case II, the MTE is a three-segmented finger-like configuration. The points are outlined to represent the morphing trailing edge starting location, the length of the first segment and the length of the second segment. The morphing location was kept at x/c = 0.60, so that enough space was provided for the actuation mechanism, and which is where its upper and lower limits were given as constraints. The upper bounds and the lower bounds were chosen to prevent the generation of morphing airfoil shapes with no likelihood of having very good aerodynamic efficiency. XFoil was used as an aerodynamic solver and therefore as a powerful tool that allowed obtaining fast convergence solutions. The results were later validated by the use of a RANS models in Ansys/Fluent solver.

Figure 19a illustrates the initial (reference) versus the final (optimized or morphed) airfoil shape, while Figure 19b illustrates the convergence trend for the minimum drag optimization function for an optimized airfoil shape with MTE at an angle of attack of 50. It is seen that the graph in Figure 19b trends to a constant after some iteration steps, which shows convergence to its optimal solution.

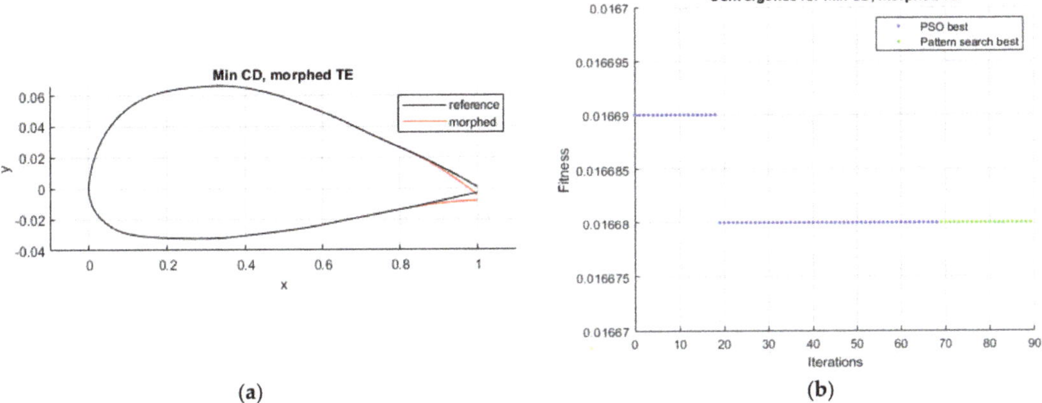

(a)

(b)

Figure 19. Optimized airfoil and convergence history for drag minimization function. (**a**) the initial (reference) versus the final (optimized or mor-phed) airfoil shape, (**b**) the convergence trend for the minimum drag optimization function.

The first set of results obtained for MTE for different flight conditions are shown in Figure 20a–d. The lift coefficients are presented for the three different flight conditions in the two different cases of Opt. Case I and Opt. Case II. The lift coefficients of the optimized airfoil design have higher values than those of the baseline airfoil; therefore, the optimized shapes produce an improvement of the lift coefficient with respect to the baseline airfoil of up to almost 26%, as seen in Figure 20c for Opt. Case I. In addition, an increase of 8% is obtained for the maximum lift coefficient for the optimized airfoils with respect to the baseline airfoil.

Figure 21 presents the evaluation of the lift versus drag variation (polar) of the baseline airfoil and the optimized airfoils for different flight conditions. The increased aerodynamic performance of optimized airfoils for all three flight conditions can be seen in Figure 21a–d.

The pressure coefficients of the MTE airfoil cases at three different flight conditions are shown in Figure 22. It can be observed that the suction peaks of the optimized airfoils are higher than those of the baseline airfoil. The MTE airfoil results expressed in terms of "effective camber change" lead to a significant increase in the negative pressure value at the trailing edge surface. The low performance of the baseline airfoil compared to that of the optimized airfoils can be related to the unfavorable pressure gradient on its upper surface near the leading edge, thereby causing an earlier laminar–turbulent transition of the boundary layer by slightly increasing the drag.

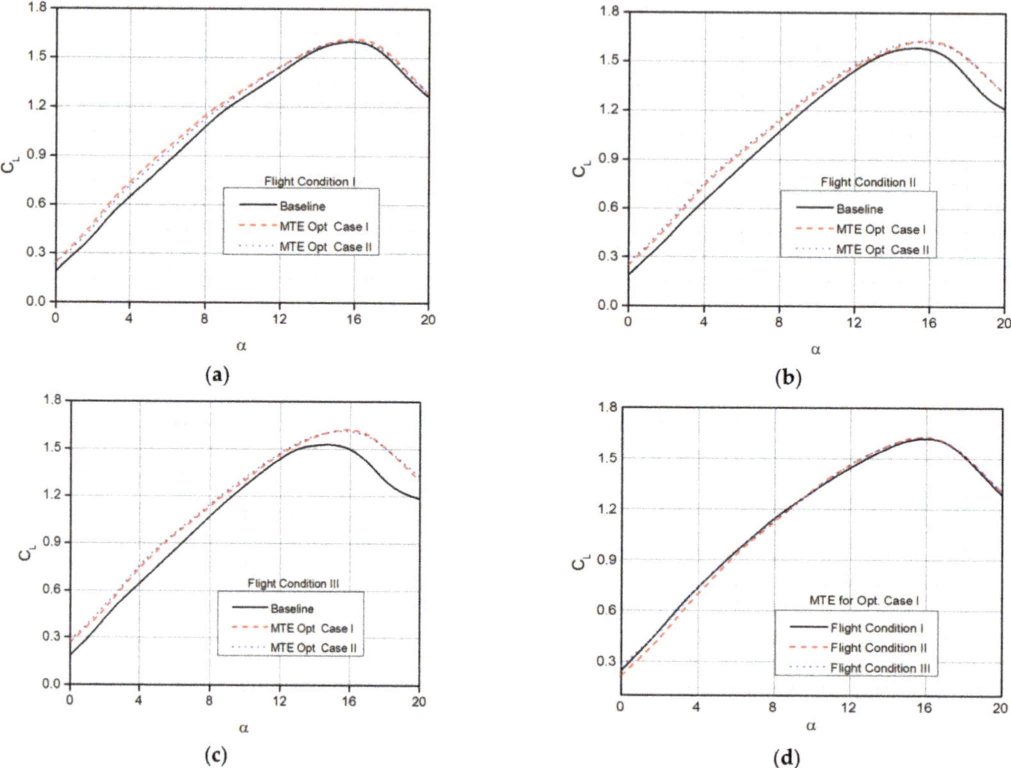

Figure 20. Lift coefficients versus angle of attack at two optimized design configurations; (**a**) the variations of lift coefficients with the angles of attack at Flight Condition I (**b**) the variations of lift coefficients with the angles of attack at Flight Condition II; (**c**) the variations of lift coefficients with the angles of attack at Flight Condition III; (**d**) the lift coeffi-cient variations for all three flight conditions for Opt. Case I.

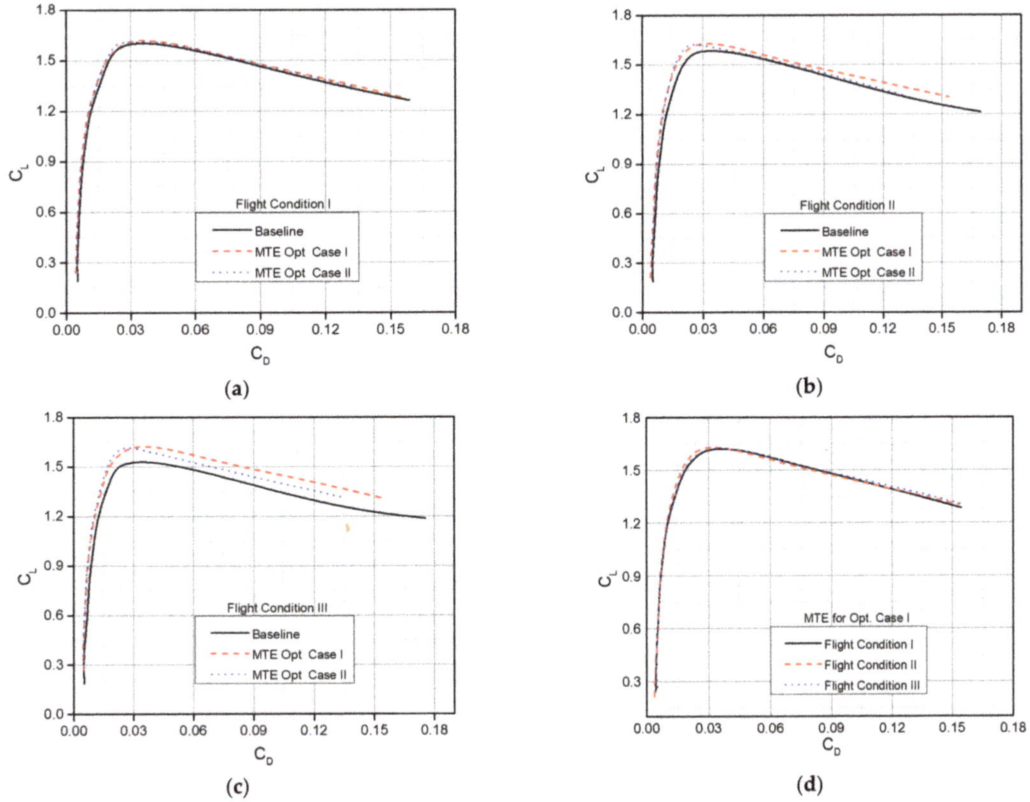

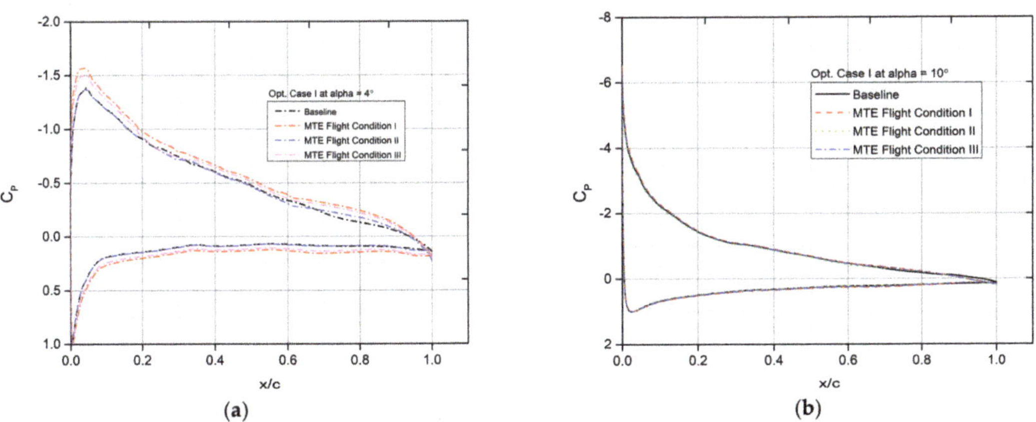

Figure 21. Lift coefficients versus drag coefficients for various optimized design configurations. (**a**) the drag polar at Flight Condition I; (**b**) drag polar at Flight Condition II; (**c**) drag polar at Flight Condition III; (**d**) drag polar for all three flight conditions for Opt. Case I.

Figure 22. Pressure coefficients for different MTE design configurations. (**a**) at angles of attack of 4°; (**b**) at angles of attack of 10°.

Figure 23a illustrates the initial (reference) versus the final (optimized or morphed) airfoil shape, while Figure 23b presents the convergence graph of the aerodynamic endurance optimization for the given MTE optimized airfoil while utilizing the PSO algorithm with pattern search. Figure 23b shows that the algorithm performs well and improves the airfoil shape to maximize the endurance.

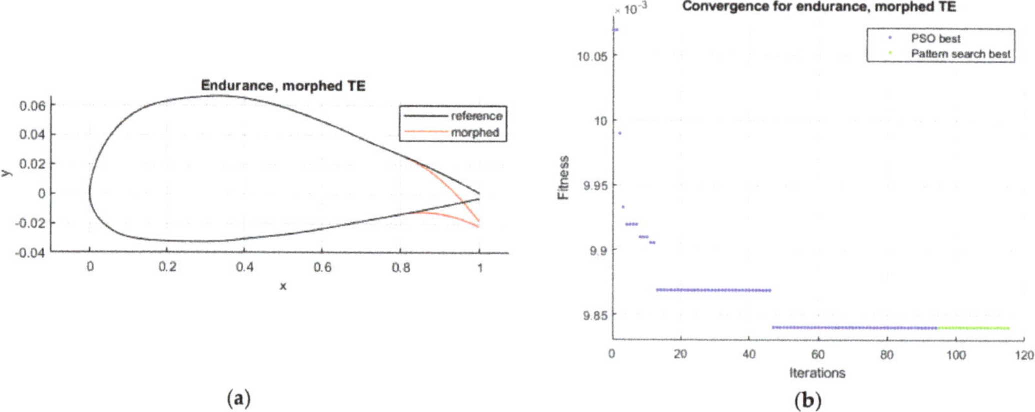

(a)

(b)

Figure 23. Optimized airfoil and convergence history for endurance maximization function. (**a**) the initial (reference) versus the final (optimized or morphed) airfoil shape; (**b**) the convergence graph of the aerodynamic endurance optimization.

The optimization process using the endurance maximization objective function led to improved endurance performance in the optimized airfoils with respect to the endurance performance of the baseline airfoil. The performance measured in terms of $C_L^{3/2}/C_D$ variation with C_L is shown for three different flight conditions in Figure 24. The endurance performance of each optimized airfoil shows a higher value in the MTE Opt. Case I with respect to its baseline airfoil configurations. For the same endurance performance defined by the $C_L^{3/2}/C_D$ ratio, the lift coefficients were found to be 1.17 for the baseline airfoil and 1.45 for the MTE airfoil configuration for the optimized flight condition II, respectively. Furthermore, as seen in Figure 24, the efficiency of $C_L^{3/2}/C_D$ increased by approximately 10.25% for flight condition III; this finding indicates the improved endurance performance of the UAS-S45 airfoil.

The difference between the results obtained by the two objective functions of drag minimization and endurance maximization to obtain optimal airfoil configurations was determined. The optimized airfoil geometrical shapes were obtained for drag minimization and endurance maximization as shown in Figure 25. The main difference observed in the airfoil geometries, as shown in Figure 25, is that the trailing edge obtained with the drag minimization objective function gave a smaller deflection than the baseline airfoil. Likewise, the endurance maximization objective function requires higher airfoil deflection with a continuous trailing edge.

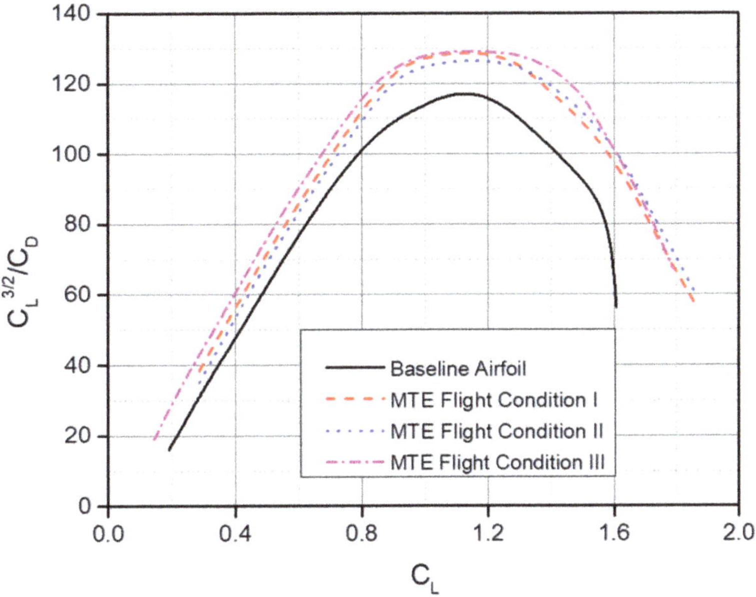

Figure 24. Comparison of the endurance performance for baseline and MTE airfoils.

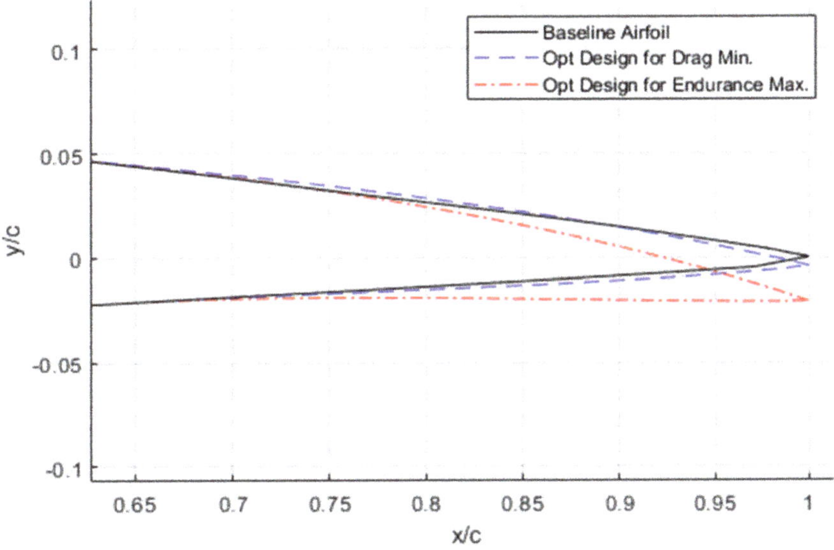

Figure 25. Comparison of MTE airfoil shapes based on different performance objectives.

The contour plots of the velocity magnitude and of the turbulent kinetic energy are visualized for the baseline airfoil and for the airfoils optimized by the Transition ($\gamma - Re_\theta$) SST turbulence model in Figure 26. These contour plots reveal that for a given MTE deflection, larger TE separation regions are found at 4° angles of attack than at 10° angles of attack. Similarly, the vertex of turbulent kinetic energy (TKE) originating from the MTE at an angle of attack of 40° has more strength than the vertex an the angle of attack of 10°.

The TKE contour plot of the baseline airfoil at an angle of attack of 4° is approximately comparable to the TKE contour plot of the MTE airfoil at an angle of attack of 10°. Therefore, the MTE-optimized airfoils can provide increased aerodynamics performance with respect to the performance of the original airfoils.

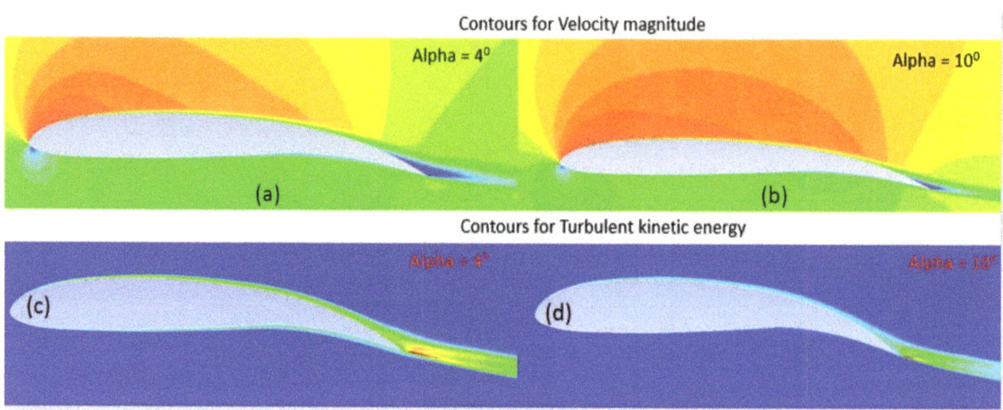

Figure 26. Velocity magnitude and Turbulent Kinetic Energy (TKE) contour plots. (**a**) Velocity contour lines at low angles of attack such as 4°; (**b**) Velocity contour lines at low angles of attack such as 10°; (**c**) the turbulent intensity at a low angle of attack of 4°; (**d**) the turbulent intensity at a low angle of attack of 10°.

4. Conclusions

This study was conducted to perform aerodynamic optimization of Morphing Trailing Edge (MTE) and Droop Nose Leading Edge (DNLE) airfoils for the UAS-S45 at different flight conditions. A new hybrid optimization technique was chosen by coupling the Particle Swarm Optimization (PSO) with the Pattern Search (PS) algorithms. The optimization function was designed to minimize drag with respect to given constraints such as airfoil lower and upper bounds, as well as to increase endurance in given flight conditions. The Bezier–PARSEC (BP) technique was used to parameterize the baseline airfoil shape, as well as to obtain its various optimized morphing configurations by using different constraints on the morphing of both the Morphing Trailing Edge (MTE) and the Droop Nose Leading Edge (DNLE) of the wing.

Within the aerodynamics optimization framework, the low-fidelity solver XFoil and the high-fidelity CFD solver Ansys Fluent were used. The results obtained using both solvers were compared for their validation. Specifically, the flow transition was predicted using Menter's Transition $(\gamma - Re_{\theta})$ SST turbulence model. In addition, the optimization framework was done from an aerodynamics perspective, and therefore, no structural studies were taken into consideration.

Both DNLE and MTE airfoil optimizations were aimed to increase the aerodynamic performance of the UAS-S45 for a wide range of angles of attack. For each of the three considered flight conditions, enhanced aerodynamic efficiency was obtained by optimizing the morphing airfoil design with respect to its UAS-S45 baseline airfoil. The optimization results have shown an increase of lift coefficients in DNLE airfoils until the stall angle of attack and thus a delay in the stall angle. An improvement in the lift coefficient was produced for the DNLE airfoils with respect to the baseline airfoil of up to 21%. In addition, an increase of 9.6% was obtained for the maximum lift coefficient, and the stall angle was also delayed by 3 degrees. Similarly, the aerodynamic performance showed a significant improvement for the MTE configurations. An increase in the maximum lift coefficient of up to 8.13% and of the efficiency of $C_L{}^{3/2}/C_D$ by 10.25% was obtained, thus indicating an increased endurance performance for the MTE airfoils of the UAS-S45. The delay of the leading edge separation by use of DNLE airfoils was another interesting result.

The flight conditions were chosen from the UAS-S45 manufacturer Hydra Technology user manual. These conditions do not cover aircraft maneuvers or gust responses. UAS-S45 maneuvers and gust response studies could be done in the future. Other objectives of this morphing optimization study will include the three-dimensional analysis of a UAS-S45 wing with a combined droop nose leading edge (DNLE) and continuous morphing trailing edge (MTE). The improvements at the performance level, such as the typical fuel savings with the optimized morphing concepts for a given flight mission, will also be studied in the future. Based on the aero-structural studies, various configurations will be analyzed, and the internal actuation mechanism will be implemented. Wind tunnel and flight tests could be further performed to validate the optimized UAS-S45 model.

Author Contributions: Conceptualization; methodology, M.B.; software, S.L.-M., and M.B.; validation, M.B.; investigation, M.B., and S.L.-M.; writing—original draft preparation, M.B.; writing—review and editing, R.M.B.; visualization, M.B., S.L.-M. and R.M.B.; supervision, R.M.B.; and T.W.; funding acquisition, R.M.B. All authors have read and agreed to the published version of the manuscript.

Funding: This research received no external funding.

Institutional Review Board Statement: Not applicable.

Informed Consent Statement: Not applicable.

Data Availability Statement: The data presented in this study are available on request from the corresponding author.

Acknowledgments: Special thanks are due to the Natural Sciences and Engineering Research Council of Canada (NSERC) for the Canada Research Chair Tier 1 in Aircraft Modelling and Simulation Technologies funding. We would also like to thank Odette Lacasse for her support at the ETS, as well as to Hydra Technologies' team members Carlos Ruiz, Eduardo Yakin and Alvaro Gutierrez Prado in Mexico.

Conflicts of Interest: The authors declare no conflict of interest.

Nomenclature

c_l	Lift force per unit span
c_d	Drag force per unit span
C_L	Lift coefficient
C_D	Drag coefficient
c	Chord
c_p	Specific fuel consumption
C_f	Skin friction coefficient
D	Drag force
D_k	Destruction term (Turbulent kinetic energy)
D_ω	Destruction term (Specific Dissipation Rate)
dz_{te}	Trailing edge thickness
e	Non-dimensional Oswald efficiency number
E	Endurance
E_a	Aerodynamic Endurance Efficiency
E_γ	Source terms (Vortex)
f	Function
H	Shape factor
k	Turbulent kinetic energy
L	Lift force

M	Mach number
P_k	Production term (kinetic energy)
r_{le}	Leading edge radius
Re_θ	Reynolds number with respect to the momentum thickness
S	Wing planform area
U	Boundary layer edge velocity
v_i^n	Velocity vector
V_∞	Free stream velocity
W_i	Initial weight of the aircraft
W_f	Final weight of the aircraft
x_i^n	Position vector
x_c	Upper crest position in horizontal coordinates
y_c	Upper crest position in vertical coordinates
z_{te}	Trailing edge offset in the vertical direction

Greek Symbols

α	Angle of attack
α_{te}	Trailing edge direction
β_{te}	Trailing edge wedge angle
η_{pr}	Propeller efficiency
ρ	Free stream density
∂	Differential operator
θ	Momentum thickness
ζ	Streamwise coordinate
μ	Dynamic viscosity coefficient
σ	Source
ω	Specific Dissipation Rate
γ	Vortex

References

1. BBC News, Q. *ATAG Beginner's Guide to Aviation Efficiency*; The Intergovernmental Panel on Climate Change (IPCC): Geneva, Switzerland, 2019.
2. Gerhards, R.; Szodruch, J. Industrial Perspectives of Drag Reduction Technologies. In *Aerodynamic Drag Reduction Technologies*; Springer: Berlin/Heidelberg, Germany, 2001; pp. 259–266.
3. Valasek, J. *Morphing Aerospace Vehicles and Structures*; John Wiley & Sons: Hoboken, NJ, USA, 2012; Volume 57.
4. Botez, R. Morphing wing, UAV and aircraft multidisciplinary studies at the Laboratory of Applied Research in Active Controls, Avionics and AeroServoElasticity LARCASE. *Aerosp. Lab* **2018**, 1–11. [CrossRef]
5. Barbarino, S.; Bilgen, O.; Ajaj, R.M.; Friswell, M.I.; Inman, D.J. A review of morphing aircraft. *J. Intell. Mater. Syst. Struct.* **2011**, *22*, 823–877. [CrossRef]
6. Ameduri, S.; Concilio, A. Morphing wings review: Aims, challenges, and current open issues of a technology. *Proc. Inst. Mech. Eng. Part C J. Mech. Eng. Sci.* **2020**. [CrossRef]
7. Communier, D.; Botez, R.M.; Wong, T. Design and Validation of a New Morphing Camber System by Testing in the Price—Païdoussis Subsonic Wind Tunnel. *Aerospace* **2020**, *7*, 23. [CrossRef]
8. Botez, R.; Koreanschi, A.; Sugar-Gabor, O.; Mebarki, Y.; Mamou, M.; Tondji, Y.; Guezguez, M.; Tchatchueng Kammegne, J.; Grigorie, L.; Sandu, D. Numerical and experimental transition results evaluation for a morphing wing and aileron system. *Aeronaut. J.* **2018**, *122*, 747–784. [CrossRef]
9. Ameduri, S.; Galasso, B.; Ciminello, M.; Concilio, A. Shape memory alloys compact actuators for aerodynamic surfaces twisting. In Proceedings of the AIAA Scitech 2020 Forum, Orlando, FL, USA, 6–10 January 2020; p. 1299.
10. Marino, M.; Sabatini, R. Advanced lightweight aircraft design configurations for green operations. In Proceedings of the Practical Responses to Climate Change Conference, Melbourne, Australia, 27 November 2014; p. 207.
11. Noviello, M.C.; Dimino, I.; Amoroso, F.; Pecora, R. Aeroelastic Assessments and Functional Hazard Analysis of a Regional Aircraft Equipped with Morphing Winglets. *Aerospace* **2019**, *6*, 104. [CrossRef]
12. Dimino, I.; Pecora, R.; Arena, M. Aircraft morphing systems: Elasticity of selected components and modelling issues. In Proceedings of the Active and Passive Smart Structures and Integrated Systems XIV, 19 May 2020; p. 113760M.
13. Pantelakis, S.; Horst, P.; Kintscher, M.; Wiedemann, M.; Monner, H.P.; Heintze, O.; Kühn, T. Design of a smart leading edge device for low speed wind tunnel tests in the European project SADE. *Int. J. Struct. Integr.* **2011**, *2*, 383–405.

14. Carossa, G.M.; Ricci, S.; De Gaspari, A.; Liauzun, C.; Dumont, A.; Steinbuch, M. Adaptive Trailing Edge: Specifications, Aerodynamics, and Exploitation. In *Smart Intelligent Aircraft Structures (SARISTU)*; Springer: Berlin/Heidelberg, Germany, 2016; pp. 143–158.
15. Li, Y.; Wang, X.; Zhang, D. Control strategies for aircraft airframe noise reduction. *Chin. J. Aeronaut.* **2013**, *26*, 249–260. [CrossRef]
16. Abbas, A.; De Vicente, J.; Valero, E. Aerodynamic technologies to improve aircraft performance. *Aerosp. Sci. Technol.* **2013**, *28*, 100–132. [CrossRef]
17. Arena, M.; Nagel, C.; Pecora, R.; Schorsch, O.; Dimino, I. Static and Dynamic Performance of a Morphing Trailing Edge Concept with High-Damping Elastomeric Skin. *Aerospace* **2019**, *6*, 22. [CrossRef]
18. Perry III, B.; Cole, S.R.; Miller, G.D. Summary of an active flexible wing program. *J. Aircr.* **1995**, *32*, 10–15. [CrossRef]
19. DeCamp, R.; HARDY, R. Mission adaptive wing research programme. *Aircr. Eng. Aerosp. Technol.* **1981**, *53*, 10–11. [CrossRef]
20. Li, D.; Zhao, S.; Da Ronch, A.; Xiang, J.; Drofelnik, J.; Li, Y.; Zhang, L.; Wu, Y.; Kintscher, M.; Monner, H.P. A review of modelling and analysis of morphing wings. *Prog. Aerosp. Sci.* **2018**, *100*, 46–62. [CrossRef]
21. Li, D.; Guo, S.; Xiang, J. Modeling and nonlinear aeroelastic analysis of a wing with morphing trailing edge. *Proc. Inst. Mech. Eng. Part G J. Aerosp. Eng.* **2013**, *227*, 619–631. [CrossRef]
22. Pecora, R.; Magnifico, M.; Amoroso, F.; Monaco, E. Multi-parametric flutter analysis of a morphing wing trailing edge. *Aeronaut. J.* **2014**, *118*, 1063–1078. [CrossRef]
23. Su, W.; Song, W. A real-time hybrid aeroelastic simulation platform for flexible wings. *Aerosp. Sci. Technol.* **2019**, *95*, 105513. [CrossRef]
24. Hang, X.; Su, W.; Fei, Q.; Jiang, D. Analytical sensitivity analysis of flexible aircraft with the unsteady vortex-lattice aerodynamic theory. *Aerosp. Sci. Technol.* **2020**, *99*, 105612. [CrossRef]
25. Zhang, J.; Shaw, A.D.; Wang, C.; Gu, H.; Amoozgar, M.; Friswell, M.I.; Woods, B.K. Aeroelastic model and analysis of an active camber morphing wing. *Aerosp. Sci. Technol.* **2021**, *106534*. [CrossRef]
26. Berci, M.; Gaskell, P.H.; Hewson, R.W.; Toropov, V.V. A semi-analytical model for the combined aeroelastic behaviour and gust response of a flexible aerofoil. *J. Fluids Struct.* **2013**, *38*, 3–21. [CrossRef]
27. Murua, J.; Palacios, R.; Peiró, J. Camber effects in the dynamic aeroelasticity of compliant airfoils. *J. Fluids Struct.* **2010**, *26*, 527–543. [CrossRef]
28. Cook, J.R.; Smith, M.J. Stability of Aeroelastic Airfoils with Camber Flexibility. *J. Aircr.* **2014**, *51*, 2024–2027. [CrossRef]
29. Sanders, B.; Crowe, R.; Garcia, E. Defense advanced research projects agency–Smart materials and structures demonstration program overview. *J. Intell. Mater. Syst. Struct.* **2004**, *15*, 227–233. [CrossRef]
30. Pendleton, E.W.; Bessette, D.; Field, P.B.; Miller, G.D.; Griffin, K.E. Active aeroelastic wing flight research program: Technical program and model analytical development. *J. Aircr.* **2000**, *37*, 554–561. [CrossRef]
31. Fortin, F. Shape optimization of a stretchable drooping leading edge. In Proceedings of the AIAA Scitech 2019 Forum, San Diego, CA, USA, 7–11 January 2019; p. 2352.
32. Koreanschi, A.; Oliviu, S.G.; Acotto, J.; Botez, R.M.; Mamou, M.; Mebarki, Y. A genetic algorithm optimization method for a morphing wing tip demonstrator validated using infra red experimental data. In Proceedings of the 34th AIAA Applied Aerodynamics Conference, Washington, DC, USA, 13–17 June 2016; p. 4037.
33. Koreanschi, A.; Sugar-Gabor, O.; Botez, R.M. Drag optimisation of a wing equipped with a morphing upper surface. *Aeronaut. J.* **2016**, *120*, 473–493. [CrossRef]
34. Arena, M.; Palumbo, R.; Pecora, R.; Amoroso, F.; Amendola, G.; Dimino, I. Flutter clearance investigation of camber-morphing aileron tailored for a regional aircraft. *J. Aerosp. Eng.* **2019**, *32*, 04018146. [CrossRef]
35. Oliviu, S.G.; Koreanschi, A.; Botez, R.M.; Mamou, M.; Mebarki, Y. Analysis of the aerodynamic performance of a morphing wing-tip demonstrator using a novel nonlinear vortex lattice method. In Proceedings of the 34th AIAA Applied Aerodynamics Conference, Washington, DC, USA, 13–17 June 2016; p. 4036.
36. Frota, J. NACRE novel aircraft concepts. *Aeronaut. J.* **2010**, *114*, 399–404. [CrossRef]
37. Schweiger, J.; Suleman, A.; Kuzmina, S.; Chedrik, V. MDO concepts for an European research project on active aeroelastic aircraft. In Proceedings of the 9th AIAA/ISSMO Symposium on Multidisciplinary Analysis and Optimization, Atlanta, GA, USA, 4–6 September 2002; p. 5403.
38. Sawyers, D. AWIATOR Project Perspectives: Passive Flow Control on Civil Aircraft Flaps Using Sub-Boundary Layer Vortex Generators. In Proceedings of the KATnet II Separation Control Workshop, San Francisco, CA, USA, 5–8 June 2006.
39. Monner, H.; Kintscher, M.; Lorkowski, T.; Storm, S. Design of a smart droop nose as leading edge high lift system for transportation aircrafts. In Proceedings of the 50th AIAA/ASME/ASCE/AHS/ASC Structures, Structural Dynamics, and Materials Conference 17th AIAA/ASME/AHS Adaptive Structures Conference 11th AIAA No, Palm Springs, CA, USA, 4–7 May 2009; p. 2128.
40. König, J.; Hellstrom, T. The Clean Sky "Smart Fixed Wing Aircraft Integrated Technology Demonstrator": Technology targets and project status. In Proceedings of the 27th International Congress of the Aeronautical Science, Nice, France, 19–24 September 2010; pp. 5101–5110.
41. Wölcken, P.C.; Papadopoulos, M. *Smart Intelligent Aircraft Structures (SARISTU): Proceedings of the Final Project Conference*; Springer: Berlin/Heidelberg, Germany, 2015.
42. Arena, M.; Concilio, A.; Pecora, R. Aero-servo-elastic design of a morphing wing trailing edge system for enhanced cruise performance. *Aerosp. Sci. Technol.* **2019**, *86*, 215–235. [CrossRef]

43. Suleman, A.; Vale, J.; Afonso, F.; Lau, F.; Ricci, S.; De Gaspari, A.; Riccobene, L.; Cavagna, L.; Cooper, J.; Wales, C. Novel air vehicle configurations: From fluttering wings to morphing flight. In Proceedings of the World Congress on Computational Mechanics (WCCM XI), Barcelona, Spain, 20–25 July 2014; pp. 20–25.

44. Ameduri, S.; Dimino, I.; Pecora, R.; Ricci, S. AIRGREEN2-Clean Sky 2 Programme: Adaptive Wing Technology Maturation, Challenges and Perspectives. In Proceedings of the ASME Conference on Smart Materials, Adaptive Structures and Intelligent Systems, San Antonio, TX, USA, 10–12 September 2018; Volume 51944, p. V001T04A023.

45. Grigorie, T.L.; Botez, R.M.; Popov, A.V.; Mamou, M.; Mébarki, Y. A hybrid fuzzy logic proportional-integral-derivative and conventional on-off controller for morphing wing actuation using shape memory alloy Part 1: Morphing system mechanisms and controller architecture design. *Aeronaut. J.* **2012**, *116*, 433–449. [CrossRef]

46. Popov, A.V.; Grigorie, T.L.; Botez, R.M.; Mébarki, Y.; Mamou, M. Modeling and testing of a morphing wing in open-loop architecture. *J. Aircr.* **2010**, *47*, 917–923. [CrossRef]

47. Skinner, S.N.; Zare-Behtash, H. State-of-the-art in aerodynamic shape optimisation methods. *Appl. Soft Comput.* **2018**, *62*, 933–962. [CrossRef]

48. Hicken, J. *Efficient Algorithms for Future Aircraft Design: CONTRIBUTIONS to Aerodynamic Shape Optimization*; University of Toronto: Toronto, ON, Canada, 2009.

49. Safari, A.; Lemu, H.G.; Jafari, S.; Assadi, M. A comparative analysis of nature-inspired optimization approaches to 2D geometric modelling for turbomachinery applications. *Math. Probl. Eng.* **2013**, *2013*, 716237. [CrossRef]

50. Zhang, J.; Xia, P. An improved PSO algorithm for parameter identification of nonlinear dynamic hysteretic models. *J. Sound Vib.* **2017**, *389*, 153–167. [CrossRef]

51. Chugh, T.; Sun, C.; Wang, H.; Jin, Y. Surrogate-assisted evolutionary optimization of large problems. In *High-Performance Simulation-Based Optimization*; Springer: Berlin/Heidelberg, Germany, 2020; pp. 165–187.

52. Lee, D.-S.; Gonzalez, L.F.; Periaux, J.; Srinivas, K. Robust design optimisation using multi-objective evolutionary algorithms. *Comput. Fluids* **2008**, *37*, 565–583. [CrossRef]

53. Martinelli, L.; Jameson, A. Computational aerodynamics: Solvers and shape optimization. *J. Heat Transf.* **2013**, *135*, 011002. [CrossRef]

54. Jameson, A.; Shankaran, S.; Martinelli, L. Continuous adjoint method for unstructured grids. *Aiaa J.* **2008**, *46*, 1226–1239. [CrossRef]

55. Lyu, Z.; Kenway, G.K.; Martins, J.R. Aerodynamic shape optimization investigations of the common research model wing benchmark. *Aiaa J.* **2014**, *53*, 968–985. [CrossRef]

56. Gabor, O.S.; Simon, A.; Koreanschi, A.; Botez, R.M. Aerodynamic performance improvement of the UAS-S4 Éhecatl morphing airfoil using novel optimization techniques. *Proc. Inst. Mech. Eng. Part G J. Aerosp. Eng.* **2016**, *230*, 1164–1180. [CrossRef]

57. Hashimoto, A.; Obayashi, S.; Jeong, S. Aerodynamic optimization of high-wing configuration for near future aircraft. In Proceedings of the 10th AIAA Multidisciplinary Design Optimization Conference, National Harbor, MD, USA, 13–17 January 2014; p. 0291.

58. Ganguli, R.; Rajagopal, S. Multidisciplinary design optimization of an UAV wing using kriging based multi-objective genetic algorithm. In Proceedings of the 50th AIAA/ASME/ASCE/AHS/ASC Structures, Structural Dynamics, and Materials Conference 17th AIAA/ASME/AHS Adaptive Structures Conference 11th AIAA No, Palm Springs, CA, USA, 4–7 May 2009; p. 2219.

59. Fincham, J.; Friswell, M. Aerodynamic optimisation of a camber morphing aerofoil. *Aerosp. Sci. Technol.* **2015**, *43*, 245–255. [CrossRef]

60. Murugan, S.; Woods, B.; Friswell, M. Hierarchical modeling and optimization of camber morphing airfoil. *Aerosp. Sci. Technol.* **2015**, *42*, 31–38. [CrossRef]

61. Albuquerque, P.F.G.L.F. Mission-Based Multidisciplinary Design Optimization Methodologies for Unmanned Aerial Vehicles with Morphing Technologies. Master's Thesis, Universidade Da Beira Interior, Covilha, Portugal, 2017; p. 10951767.

62. Khurana, M.; Winarto, H.; Sinha, A. Application of swarm approach and artificial neural networks for airfoil shape optimization. In Proceedings of the 12th AIAA/ISSMO Multidisciplinary Analysis and Optimization Conference, Victoria, BC, Canada, 10–12 September 2008; p. 5954.

63. Kao, J.Y.; Clark, D.L.; White, T.; Reich, G.W.; Burton, S. Conceptual Multidisciplinary Design and Optimization of Morphing Aircraft. In Proceedings of the AIAA Scitech 2019 Forum, San Diego, CA, USA, 7–11 January 2019; p. 0175.

64. Magrini, A.; Benini, E.; Ponza, R.; Wang, C.; Khodaparast, H.H.; Friswell, M.I.; Landersheim, V.; Laveuve, D.; Contell Asins, C. Comparison of constrained parameterisation strategies for aerodynamic optimisation of morphing leading edge airfoil. *Aerospace* **2019**, *6*, 31. [CrossRef]

65. Gong, C.; Ma, B.-F. Aerodynamic evaluation of an unmanned aerial vehicle with variable sweep and span. *Proc. Inst. Mech. Eng. Part G J. Aerosp. Eng.* **2019**, *233*, 0954410019836907. [CrossRef]

66. Buckley, H.P.; Zhou, B.Y.; Zingg, D.W. Airfoil optimization using practical aerodynamic design requirements. *J. Aircr.* **2010**, *47*, 1707–1719. [CrossRef]

67. Segui, M.; Botez, R.M. Evaluation of the impact of morphing horizontal tail design of the UAS-S45 performances. In Proceedings of the Canadian Aeronautics and Space Institute (CASI), Conference, Laval, QC, Canada, 14–16 May 2019.

68. Okrent, J. An Integrated Method for Airfoil Optimization. Master's Thesis, Lehigh University, Bethlehem, PA, USA, 2017.

69. Khurana, M.; Winarto, H.; Sinha, A. Airfoil geometry parameterization through shape optimizer and computational fluid dynamics. In Proceedings of the 46th AIAA Aerospace Sciences Meeting and Exhibit, Reno, NV, USA, 7–10 January 2008; p. 295.
70. Trad, M.H.; Segui, M.; Botez, R.M. Airfoils Generation Using Neural Networks, CST Curves and Aerodynamic Coefficients. In Proceedings of the AIAA AVIATION FORUM, 15–19 June 2020; p. 2773.
71. Derksen, R.; Rogalsky, T. Optimum aerofoil parameterization for aerodynamic design. *Comput. Aided Optim. Des. Eng. Xi* **2009**, *106*, 197–206.
72. Derksen, R.; Rogalsky, T. Bezier-PARSEC: An optimized aerofoil parameterization for design. *Adv. Eng. Softw.* **2010**, *41*, 923–930. [CrossRef]
73. Kharal, A.; Saleem, A. Neural networks based airfoil generation for a given Cp using Bezier–PARSEC parameterization. *Aerosp. Sci. Technol.* **2012**, *23*, 330–344. [CrossRef]
74. Khurana, M. Development and Application of an Optimisation Architecture with Adaptive Swarm Algorithm for Airfoil Aerodynamic Design. Ph.D. Thesis, RMIT University, Melbourne, VIC, Australia, 2011.
75. Drela, M.; Youngren, H. XFOIL: Interactive Program for the Design and Analysis of Subsonic Isolated Airfoils. 2015. Available online: https://web.mit.edu/drela/Public/web/xfoil/ (accessed on 11 February 2021).
76. Van de Wal, H. Design of a Wing with Boundary Layer Suction. Master's Thesis, Tu Delft, Delft, The Netherlands, 2010.
77. Woods, B.K.; Fincham, J.H.; Friswell, M.I. Aerodynamic modelling of the fish bone active camber morphing concept. In Proceedings of the RAeS Applied Aerodynamics Conference, Bristol, UK, 22–24 July 2014.
78. Langtry, R.; Menter, F. Transition modeling for general CFD applications in aeronautics. In Proceedings of the 43rd AIAA Aerospace Sciences Meeting and Exhibit, Reno, NV, USA, 10–13 January 2005; p. 522.
79. Khan, S.; Grigorie, T.; Botez, R.; Mamou, M.; Mébarki, Y. Novel morphing wing actuator control-based Particle Swarm Optimisation. *Aeronaut. J.* **2020**, *124*, 55–75. [CrossRef]
80. Burnazzi, M.; Radespiel, R. Assessment of leading-edge devices for stall delay on an airfoil with active circulation control. *Ceas Aeronaut. J.* **2014**, *5*, 359–385. [CrossRef]

Article

Several Cases for the Validation of Turbulence Models Implementation

Michael D. Polewski and Paul G. A. Cizmas *

Department of Aerospace Engineering, Texas A&M University, 701 H. R. Bright Building, College Station, TX 77843-3141, USA; mpolewski@aggienetwork.com
* Correspondence: cizmas@tamu.edu; Tel.: +1-979-845-5952

Abstract: This paper presents several test cases that were used to validate the implementation of two turbulence models in the UNS3D code, an in-house code. The two turbulence models used were the Shear Stress Transport model and the Spalart–Allmaras model. These turbulence models were explored using the numerical results generated by three computational fluid dynamics codes: NASA's FUN3D and CFL3D, and UNS3D. Four cases were considered: a flat plate case, an airfoil near-wake, a backward-facing step, and a turbine cascade known as the Eleventh Standard Configuration. The numerical results were compared among themselves and against experimental data.

Keywords: computational fluid dynamics; turbulence modeling; flat plate flow; backward-facing step flow; airfoil flow; cascade flow

Citation: Polewski, M.D.; Cizmas, P.G.A. Several Cases for the Validation of Turbulence Models Implementation. *Appl. Sci.* **2021**, *11*, 3377. https://doi.org/10.3390/app11083377

Academic Editor: Ruxandra Mihaela Botez

Received: 15 March 2021
Accepted: 5 April 2021
Published: 9 April 2021

Publisher's Note: MDPI stays neutral with regard to jurisdictional claims in published maps and institutional affiliations.

1. Introduction

Computational fluid dynamics (CFD) has become an integral part of the design systems in aerospace and other engineering fields. The increase in computational power in the last two decades allows us to tackle complex, engineering relevant problems. Using today's supercomputers, it has become customary to solve transport phenomena problems with tens or hundreds of millions of grid points and generate results overnight.

As the hardware improves, the models used for predicting transport phenomena should also improve. One standing challenge for flow simulation is the turbulence modeling. As direct numerical simulation is still too computationally expensive for engineering relevant flows, the governing mass, momentum, and energy equations must be closed using a turbulence model.

The widely used $\kappa - \epsilon$ turbulence model developed by Jones and Launder [1] cannot properly capture the turbulent boundary layer up to flow separation. The first turbulence model that accurately predicted separated airfoil flows was the Johnson–King model [2]. Being an algebraic model, the Johnson–King model was not easily extensible to three-dimensional flow solvers. The $\kappa - \omega$ turbulence model proposed by Wilcox [3] improved the prediction for adverse pressure-gradient flows. In addition, the model has a simple formulation in the viscous sublayer. The $\kappa - \omega$ model, however, depends strongly on the freestream values of the specific turbulent dissipation rate, ω_f. This issues was addressed by combining the $\kappa - \omega$ and $\kappa - \epsilon$ models, so that the former models the inner region of the boundary layer and the latter models outer part. The resulting model, called the Shear Stress Transport turbulence model [4], was developed to respond to the need for accurate prediction of flows with strong adverse pressure gradients and separation. A turbulence model that is less computationally expensive than the Shear Stress Transport turbulence model is the Spalart–Allmaras model [5]. This is a one-equation model that models an eddy viscosity-like variable, $\tilde{\nu}$.

The purpose of this paper is to provide several cases that can be used to validate the implementation of the Shear Stress Transport and the Spalart–Allmaras turbulence models.

The results generated using an in-house CFD code and two NASA codes, using a couple of turbulence models, are compared with experimental data.

The next section presents the physical model and provides details on the governing equations. This is followed by a brief summary of the numerical method used to solve the governing equations. The results section presents four validation cases, ranging from a flat plate to a cascade of turbine airfoils.

2. Physical Model

The governing equations used by the CFD solvers include of the mass, momentum, and energy conservation. These equations are supplemented by the equations used for modeling the turbulence effects. This section briefly presents the integral form of the mass, momentum, and energy conservation equations, as well as the turbulence models used in this work.

2.1. Mass, Momentum and Energy Conservation Equations

For an arbitrary finite volume, Ω, bounded by a closed surface, $\partial\Omega$, a small surface, dS, with a unit normal vector, $\hat{n}$, pointing outward from the control volume, the mass conservation equation is

$$\frac{\partial}{\partial t}\int_\Omega \rho d\Omega + \oint_{\partial\Omega} \rho(\vec{v}\cdot\hat{n})dS = 0 \tag{1}$$

where ρ is the density and $\vec{v}$ is the velocity vector.

The momentum equation is

$$\frac{\partial}{\partial t}\int_\Omega \rho\vec{v}d\Omega + \oint_{d\Omega} \rho\vec{v}(\vec{v}\cdot\hat{n})dS = \oint(-p\mathbf{I}+\boldsymbol{\tau})\cdot\hat{n}dS \tag{2}$$

where the scalar p is the pressure, μ is the dynamic viscosity coefficient, $\mathbf{I}$ is the identity tensor, and $\boldsymbol{\tau}$ is the viscous stress tensor.

The energy conservation equation for a calorically perfect gas is

$$\frac{\partial}{\partial t}\int_\Omega \rho E d\Omega + \oint_{\partial\Omega} \rho H(\vec{v}\cdot\hat{n})dS = \oint_{\partial\Omega} k(\nabla T\cdot\hat{n}) + (\boldsymbol{\tau}\cdot\vec{v})\hat{n}dS \tag{3}$$

where E is the total energy, H is the total enthalpy, T is temperature, and k is thermal conductivity.

2.2. Turbulence Models

Turbulence models are essential for producing accurate results when using Reynolds averaging. They provide the closure needed for the Reynolds-averaged Navier–Stokes (RANS) equations. The following sections describe the two turbulence models used in this work: the Shear Stress Transport (SST) turbulence model and the Spalart–Allmaras (S–A) turbulence model. Both turbulence models employ the Boussinesq eddy-viscosity hypothesis, where the Favre-averaged turbulent shear stresses, τ^F, is defined by

$$\tau_{ij}^F = 2\mu_T S_{ij} - \left(\frac{2\mu_T}{3}\right)\frac{\partial\bar{v}_k}{\partial x_k}\delta_{ij} - \frac{2}{3}\bar{\rho}\kappa\delta_{ij}. \tag{4}$$

where μ_T is the turbulent eddy viscosity, S_{ij} is the strain rate tensor, κ is the kinetic energy of the turbulent fluctuations, and δ_{ij} is the Kronecker delta.

2.2.1. Shear Stress Transport Turbulence Model

The Shear Stress Transport turbulence [4] model is a two-equation turbulence model that is a combination of a high Reynolds number $\kappa - \epsilon$ turbulence model and the $\kappa - \omega$ turbulence model by Wilcox [6]. κ denotes the kinetic energy of the turbulent fluctuations, ϵ is the turbulence dissipation per unit mass, and ω is the root mean square fluctuating vorticity (or the rate of dissipation of energy in unit volume and time). The goal of combining the two models is to retain the best attributes of both models while fixing their weaknesses. For example, as the $\kappa - \omega$ model is sensitive to the freestream values of ω, the Shear Stress Transport model switches to the $\kappa - \epsilon$ model to resolve this issue. The Shear Stress Transport model is defined by the transport equations of the kinetic energy of the turbulent fluctuations, κ,

$$\frac{\partial(\rho\kappa)}{\partial t} + \frac{\partial(\rho v_j \kappa)}{\partial x_j} = \frac{\partial}{\partial x_j}\left[(\mu_L + \sigma_\kappa \mu_T)\frac{\partial\kappa}{\partial x_j}\right] + \tau_{ij}^F \frac{\partial u_i}{\partial x_j} - \beta^* \rho\omega\kappa \tag{5}$$

and of the dissipation ω,

$$\frac{\partial(\rho\omega)}{\partial t} + \frac{\partial(\rho v_j \omega)}{\partial x_j} = \frac{\partial}{\partial x_j}\left[(\mu_L + \sigma_\omega \mu_T)\frac{\partial\omega}{\partial x_j}\right] + \tau_{ij}^F \frac{\partial u_i}{\partial x_j}\frac{C_\omega \rho}{\mu_T} - \beta\rho\omega^2 +$$

$$+ 2(1 - f_1)\frac{\rho\sigma_{\omega 2}}{\omega}\frac{\partial\kappa}{\partial x_j}\frac{\partial\omega}{\partial x_j}. \tag{6}$$

f_1 is the blending function between the $\kappa - \epsilon$ and the $\kappa - \omega$ turbulence models. This blending allows the Shear Stress Transport turbulence model to switch between the two turbulence models depending on (i) the distance from the wall and (ii) the region of the turbulent boundary layer is being analyzed. Variables β^* and $\sigma_{\omega 2}$ are constant coefficients. Variables σ_ω, σ_κ, C_w, and β are non-constant coefficients that utilize the blending function, f_1 [4].

2.2.2. Spalart–Allmaras Turbulence Model

The Spalart–Allmaras turbulence model [5] is a one-equation turbulence model that solves the transport equation of an eddy viscosity-like variable, $\tilde{v}$,

$$\frac{D\tilde{v}}{Dt} = P - D + \frac{1}{\sigma}\left[\nabla \cdot ((v_L + \tilde{v})\nabla\tilde{v}) + c_{b2}(\nabla\tilde{v})^2\right] \tag{7}$$

where P and D are the production and destruction source terms,

$$P = c_{b1}(1 - f_{t2})\tilde{S}\tilde{v} \tag{8}$$

$$D = \left(c_{w1}f_w - \frac{c_{b1}}{\kappa_{SA}^2}f_{v2}\right)\left[\frac{\tilde{v}}{d}\right]^2. \tag{9}$$

Constant coefficients include σ, c_{b1}, c_{b2}, c_{w1}, and κ_{SA}. The value d is the distance to the wall. Function f_w is known as the wall function, as it is dependent on distance to wall, d. Viscous function f_{v2} depends heavily on the ratio between the molecular kinematic viscosity, v_L, and the eddy-viscosity-like variable $\tilde{v}$. The modified vorticity term, $\tilde{S}$, is a function of the magnitude of vorticity, wall distance, and viscosities [5]. The Spalart–Allmaras turbulence model is a popular turbulence model for turbomachinery flows as it was designed for wall-bounded flows. As the Spalart–Allmaras turbulence model only solves for one equation, it is also inherently less computationally expensive than Shear Stress Transport.

The original turbulence model [5] has an additional trip term, f_{t1}, used to model transition to turbulence. Allmaras et al. [7] found that the two trip terms—f_{t1} and f_{t2}—are unnecessary to simulate fully turbulent flows. The implementation used in this work does not use f_{t1} [8]. In addition, the implementation is based on the compressible version that includes an update of the modified vorticity term [7].

3. Numerical Method

The turbulence models tested herein were implemented in three RANS solvers: CFL3D, FUN3D, and UNS3D. The first two flow solvers were developed by NASA. The third flow solver was developed at Texas A&M University. As detailed information is available for the NASA codes, this section focuses on presenting the numerical method used in the UNS3D code. UNS3D stands for Unstructured, Unsteady Three-Dimensional flow solver.

3.1. Spatial Discretization

The UNS3D flow solver uses a finite volume method for its spatial discretization. The UNS3D and the NASA's FUN3D codes use a dual-mesh cell-vertex approach, while the NASA's CFL3D code uses a cell-centered approach. The cell-vertex approach has better adaptability with mixed element grids. In addition, as for large grids the number of nodes is smaller than the number of elements, the cell-vertex method is computationally more efficient than the cell-centered method.

3.1.1. Convective Flux

UNS3D has several options for calculating the convective fluxes [9]. Herein, we selected an upwind scheme developed by Roe [10] with the additional entropy correction developed by Harten [11], known as the Roe's flux-difference splitting scheme with the Harten entropy fix. The upwind scheme was chosen for its stability and ability to capture shock waves and boundary layers accurately.

3.1.2. Diffusive Flux

The diffusive flux describes flux of ϕ entering the control volume Ω due to the diffusion of ϕ. The diffusive flux is proportional to the gradient of ϕ and is described as

$$\oint_{\partial\Omega} k\rho(\nabla\phi \cdot \hat{n})dS \tag{10}$$

where k is a diffusivity coefficient, ρ is the density, and $\nabla\phi$ is the gradient of ϕ. As the gradients are stored at the nodes, a modified central scheme is used to calculate the edge-based gradients. An ordinary central scheme provides uncoupling between the local terms and the edge-based gradients. The following modification has been added to determine the edge-based gradients:

$$\nabla\phi_{ij} = \frac{1}{2}\left(\nabla\phi_i + \nabla\phi_j\right) - \left[\frac{1}{2}\left(\nabla\phi_i + \nabla\phi_j\right) \cdot \hat{e}_{ij}\right]\hat{e}_{ij} + \frac{\phi_j - \phi_i}{\vec{x}_j - \vec{x}_i}\hat{e}_{ij}. \tag{11}$$

To calculate the diffusive flux, and for second-order spatial discretization, the gradients at the nodes must be calculated. As UNS3D is an unstructured and cell-vertex dual-mesh flow solver, there are specific numerical methods that can be used to calculate the gradients. The following gradient methods are available in UNS3D: Green-Gauss, least-squares, least squares with QR decomposition, and the Weighted Essentially Non-Oscillatory method. The results presented in this paper were generated using the Least Squares with QR decomposition as the primary gradient calculation method.

3.1.3. Second-Order Spatial Discretization

The state variables and their gradients, which are located at the nodes, are required to define a piecewise, linearly-varying approximation for a continuous flow field [12]. To achieve a linear variation, the nodal values across edge "ij" are reconstructed and defined as "left" and "right" state variables.

$$\vec{Q}_L = \vec{Q}_i + \frac{1}{2}\Psi_i\left(\nabla\vec{Q}_i \cdot (\vec{x}_j - \vec{x}_i)\right)$$
$$\vec{Q}_R = \vec{Q}_j - \frac{1}{2}\Psi_j\left(\nabla\vec{Q}_j \cdot (\vec{x}_j - \vec{x}_i)\right) \tag{12}$$

where Ψ is a limiter function. The limiter functions Ψ range from 0 to 1, and prevent non-physical oscillations and spurious solutions in the vicinity of large gradients. The solution limiters accomplish this by enforcing a monotonicity preserving scheme. The solution limiters used in this work were those proposed by Venkatakrishnan [13], Carpenter [14], and Dervieux [15].

3.2. Temporal Discretization

The terms of the Navier–Stokes equations can be grouped into spatial and temporal derivatives. The spatial derivatives consist of the source terms, and the convective and diffusive flux terms. These spatial derivative terms are grouped together to form the residual $\vec{R}$, thus the Navier–Stokes equations can be written as

$$\frac{\partial}{\partial t}\vec{Q}_i(\Omega_i) = \vec{R}_i \tag{13}$$

where $\vec{Q}_i$ is the state vector, Ω_i is the control volume, and $\vec{R}_i$ is the residual at node i.

The UNS3D solver can integrate (13) using either an explicit or implicit method. The results presented in this paper were obtained using only explicit time integration. Explicit time-integration utilizes either a first-order, forward finite difference approximation for the time derivative or a second-order scheme. The second-order time integration uses a four-stage Runge–Kutta method. Implicit residual smoothing is also available in the UNS3D solver. For unsteady simulations, UNS3D utilizes either single- or dual-time stepping.

3.3. Boundary Conditions

The implementation of boundary conditions can greatly affect the accuracy and convergence of the solution. The boundary conditions in the UNS3D solver have weak implementations. Weak implementation of boundary conditions refers to the scenario when the boundary conditions are not being directly applied to the nodal values, but instead to the boundary nodes created by the dual mesh. These boundary nodes are used in determining the fluxes from the boundary, and it is in these fluxes that the boundary conditions are satisfied.

3.3.1. Flow Boundary Conditions

For the subsonic inlet, the four conditions imposed from upstream infinity are the two components of the angle of attack, the total pressure, and the total temperature. The condition imposed from the interior of the domain is the outgoing Riemann invariant, $u + 2c/(\gamma - 1)$. For subsonic outlets, only one condition is imposed from downstream infinity, which is a user-specified static pressure. The conditions enforced from the interior of the domain are the entropy; tangential velocity; and the incoming and outgoing Riemann invariants, $u + 2c/(\gamma - 1)$ and $u - 2c/(\gamma - 1)$, respectively.

The wall boundaries are either inviscid and viscous wall boundaries. All wall boundaries enforce the no-penetration condition [16]. This is the only condition for inviscid walls. For viscous flows, the boundary conditions consist of the no-penetration and no-slip conditions [17], p. 80.

Symmetry boundary conditions are used to truncate the computational domain whenever there is a plane of symmetry. Symmetry boundary conditions are also used to simulate a two-dimensional (2D) flow field with a three-dimensional (3D) flow solver, creating what is known as a quasi-3D simulation. There are two conditions that must be satisfied with symmetry boundaries: no-penetration and the zero-gradient of state variables on the planes of symmetry. For this paper, a strong implementation of the zero-gradient condition was enforced, that is, both the boundary state vectors and state vectors on the symmetry boundary had the zero-gradient condition applied [18].

3.3.2. Shear Stress Transport Boundary Conditions

The Shear Stress Transport turbulence model primarily utilizes Neumann boundary conditions. The boundary conditions used are the recommended values from the original reference for the Shear Stress Transport turbulence model [4]. These boundary conditions are listed in Table 1.

Table 1. Shear Stress Transport boundary conditions (L-length of computational domain, U_∞-velocity at upstream infinity).

Boundary	κ	ω
Wall	$\kappa = 0$	$\omega = 10\dfrac{6\nu}{\beta_1(\delta d_1)^2}$
Freestream	$\dfrac{10^{-5}U_\infty^2}{Re_L} < \kappa < \dfrac{0.1U_\infty^2}{Re_L}$	$\dfrac{U_\infty}{L} < \omega < \dfrac{10U_\infty}{L}$
Inlet	$\dfrac{10^{-5}U_\infty^2}{Re_L} < \kappa < \dfrac{0.1U_\infty^2}{Re_L}$	$\dfrac{U_\infty}{L} < \omega < \dfrac{10U_\infty}{L}$
Symmetry	$\dfrac{\partial \kappa}{\partial n} = 0.0$	$\dfrac{\partial \omega}{\partial n} = 0.0$

The symmetry boundary condition for the Shear Stress Transport turbulence model follows the same structure as the governing equations, where the zero-gradient normal is applied for κ and ω. For the freestream and inlet boundaries, Dirichlet boundaries are enforced for a specific range. The freestream values of ω are obtained from turbulent intensities [4,19]. For the outlet, the values of κ and ω are simply extrapolated. Periodic boundaries are applied to κ and ω in the same manner as to the state vectors in the full-order model.

3.3.3. Spalart–Allmaras Boundary Conditions

The Spalart–Allmaras turbulence model utilizes both the Dirichlet and Neumann boundary conditions for its working variable $\tilde{\nu}$. Dirichlet boundary conditions are applied to the inlet/freestream and wall boundaries, while Neumann boundary conditions are applied for symmetry boundaries. Table 2 lists the boundary conditions for the Spalart–Allmaras turbulence model.

Table 2. Spalart–Allmaras boundary conditions.

Boundary	Dirichlet	Neumann
Wall	-	$\tilde{\nu} = 0.0$
Freestream	-	$\dfrac{\tilde{\nu}}{\nu_L} = 3 - 5$ (fully turbulent)
Inlet	-	$\dfrac{\tilde{\nu}}{\nu_L} = 3 - 5$ (fully turbulent)
Symmetry	$\dfrac{\partial \tilde{\nu}}{\partial n} = 0.0$	-

The symmetry boundary condition for the Spalart–Allmaras turbulence model follows the same structure as that of the governing equations, where the zero-gradient in the normal direction is applied for $\tilde{\nu}$. For the freestream and inlet boundaries, a $\tilde{\nu}/\nu_L$ value between 3 and 5 was used to simulate fully turbulent flows. For tripping the laminar-to-turbulent transition, $\tilde{\nu}/\nu_L$ must be less than 1. For the outlet, the value of $\tilde{\nu}$ is extrapolated. Periodic boundaries are applied to $\tilde{\nu}$ in the same manner as the state vectors in the mass, momentum, and energy conservation equations.

4. Results

The cases used for assessing the Spalart–Allmaras turbulence model follow the NASA's turbulence modeling resource website [20]. There are two versions of the UNS3D code: a sequential version, UNS3D-SEQ, and a parallel version, UNS3D-PAR. Both the parallel and sequential versions of UNS3D implemented the Spalart–Allmaras turbulence model and underwent the same tests. The Shear Stress Transport turbulence model was also assessed using the same test cases. The results generated using UNS3D were compared against experimental data and numerical results obtained using FUN3D and CFL3D. The results of the sequential version of UNS3D used a coarser grid than that used by the parallel UNS3D and the NASA codes FUN3D and CFL3D.

4.1. Turbulent Flat Plate

The first case investigated was the turbulent flow over a flat plate. The turbulent flat plate case allowed one to validate the turbulence model against Coles' Law of the Wake [21]. The velocity profile and distance from the wall were nondimensionalized to highlight the different regions of the turbulent boundary layer such as the log-region sublayer. Both the sequential and parallel versions of UNS3D were compared against the results from NASA's FUN3D and CFL3D [20]. These results include dimensionless turbulent viscosity μ_T/μ_∞, κ, and ω contours.

The computational domain and boundary conditions are illustrated in Figure 1. Two levels of mesh refinement were used for this case: where the parallel code used the finest grid and the sequential code used the third finest grid [20]. Figure 2 shows the coarse mesh. Table 3 specifies the freestream boundary conditions for the various turbulence models and codes. The freestream turbulence boundary conditions used in NASA's FUN3D SST and CFL3D SST were above the recommended range [4], which is

$$\frac{U_\infty}{L} < \omega_\infty < 10\frac{U_\infty}{L} \qquad\qquad 10^{-5}\frac{U_\infty^2}{Re_L} < \kappa_\infty < 0.1\frac{U_\infty^2}{Re_L}.$$

The values used by NASA FUN3D correspond to

$$\omega_\infty = 125\frac{U_\infty}{L} \qquad\qquad \kappa_\infty = 1.125\frac{U_\infty^2}{Re_L}.$$

Therefore, FUN3D used turbulent freestream conditions that were approximately two orders of magnitude higher than the recommended range. UNS3D-PAR SST used values of κ and ω that were in the recommended range:

$$\omega_\infty = 5\frac{U_\infty}{L} \qquad\qquad \kappa_\infty = 0.05\frac{U_\infty^2}{Re_L}.$$

We also generated UNS3D-PAR SST results using the turbulence boundary conditions used by the NASA codes. In this case, UNS3D-PAR SST produced κ and ω contours that were approximately 10 percent smaller in magnitude than the NASA results. The results included in this paper for UNS3D-PAR SST used the turbulent boundary conditions in the recommended range, as stated in Table 3.

Table 3. Turbulent Boundary Conditions for Flat Plate.

Code	Mesh	$\frac{\tilde{v}_\infty}{v_L}$	$\omega_\infty \frac{v_L}{c_\infty^2}$	$\kappa_\infty \frac{1}{c_\infty^2}$
UNS3D-SEQ SA	$2 \times 137 \times 97$	3.0	-	-
UNS3D-PAR SA	$2 \times 545 \times 385$	3.0	-	-
FUN3D SA	$2 \times 545 \times 385$	3.0	-	-
UNS3D-SEQ SST	$2 \times 137 \times 97$	-	1×10^{-6}	9×10^{-9}
UNS3D-PAR SST	$2 \times 545 \times 385$	-	4×10^{-8}	4×10^{-10}
FUN3D SST	$2 \times 545 \times 385$	-	1×10^{-6}	9×10^{-9}

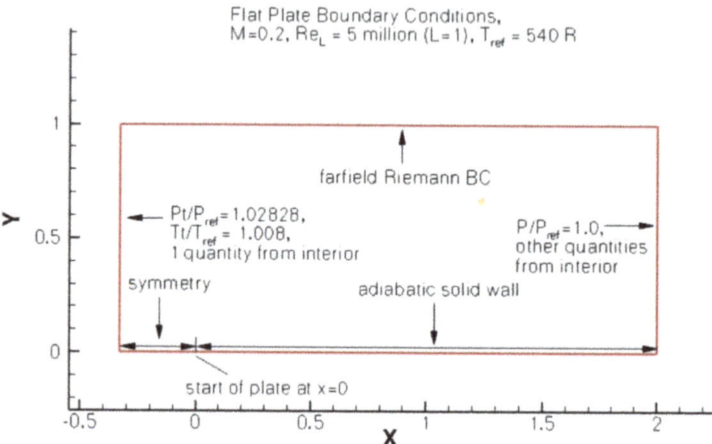

Figure 1. Flat plate boundary conditions [20].

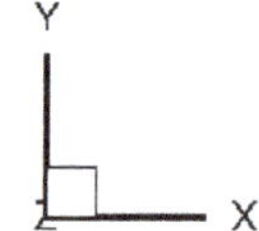

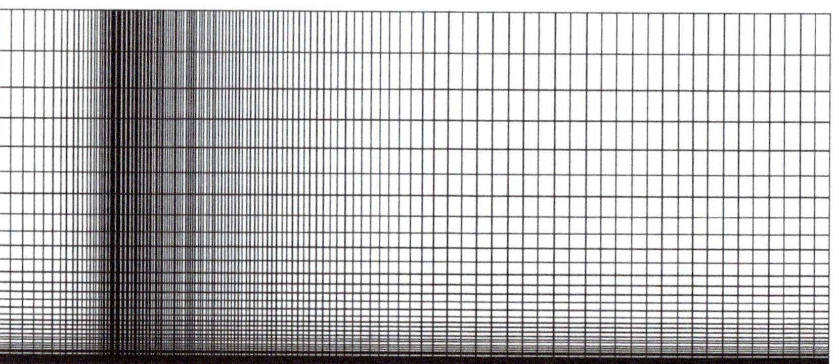

Figure 2. Flat plate coarse mesh.

Figure 3 shows a comparison of the dimensionless turbulent viscosity predicted by the UNS3D-SEQ, UNS3D-PAR, and FUN3D codes using the Spalart–Allmaras turbulence model. A good agreement was obtained among the solutions of the three codes. A similar conclusion emerged by comparing the dimensionless turbulent viscosity predicted using the Shear Stress Transport turbulence model, shown in Figure 4. The large values of the turbulent viscosity extended farther away from the flat plate in the Spalart–Allmaras model compared to the Shear Stress Transport model, as observed by contrasting Figures 3 and 4.

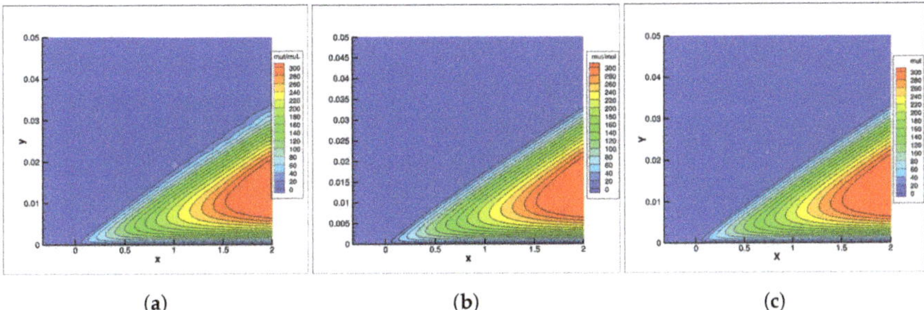

Figure 3. Contours of μ_T/μ_∞ for flat plate using Spalart–Allmaras turbulence model: (**a**) UNS3D-SEQ, (**b**) UNS3D-PAR, and (**c**) FUN3D [20].

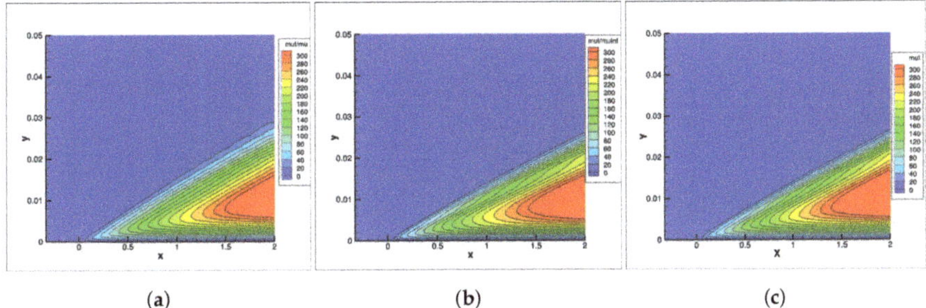

Figure 4. Contours of μ_T/μ_∞ for flat plate using Shear Stress Transport turbulence'model: (**a**) UNS3D-SEQ, (**b**) UNS3D-PAR, and (**c**) FUN3D [20].

Figure 5 shows the contours of the kinetic energy of the turbulent fluctuations, κ, produced by the Shear Stress Transport turbulence model. Similarly to the turbulent viscosity results, a good agreement is obtained among the solutions of the three codes.

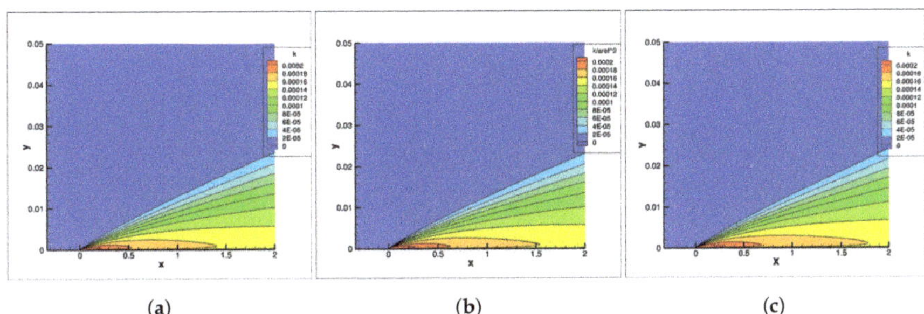

Figure 5. κ contours on flat plate using Shear Stress Transport turbulence model: (**a**) UNS3D-SEQ, (**b**) UNS3D-PAR, and (**c**) FUN3D [20].

Figure 6 shows that the main differences between the ω contours generated by UNS3D-PAR and FUN3D using the Shear Stress Transport model were located in the region near the inlet of the domain. FUN3D ω contours started at high values and dissipated as the flow went out of the domain. This difference was to be expected as the turbulent boundary conditions for FUN3D were at least one order of magnitude higher than those of the UNS3D-PAR. Overall, however, the UNS3D-PAR results were in good agreement with those of FUN3D in the region near the wall.

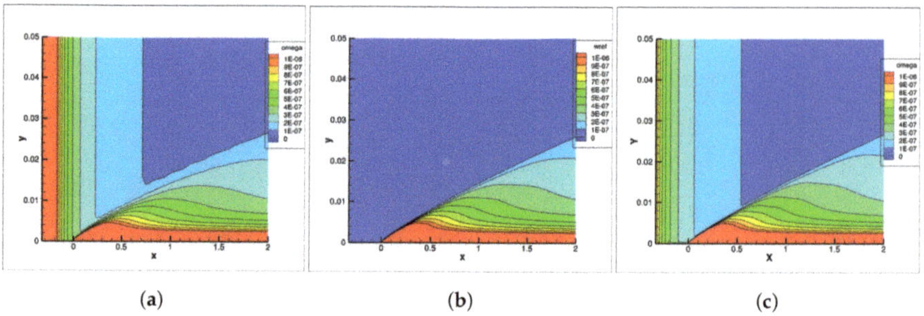

Figure 6. ω contours on flat plate using Shear Stress Transport turbulence model: (**a**) UNS3D-SEQ, (**b**) UNS3D-PAR, and (**c**) FUN3D [20].

Figure 7 shows the variation of dimensionless velocity $u^+ = u/u^*$, where u^* is the friction velocity, as a function of y^+. A good agreement is observed among the results generated by UNS3D-SEQ, UNS3D-PAR, and CFL3D codes, for both the Spalart–Allmaras and Shear Stress Transport models.

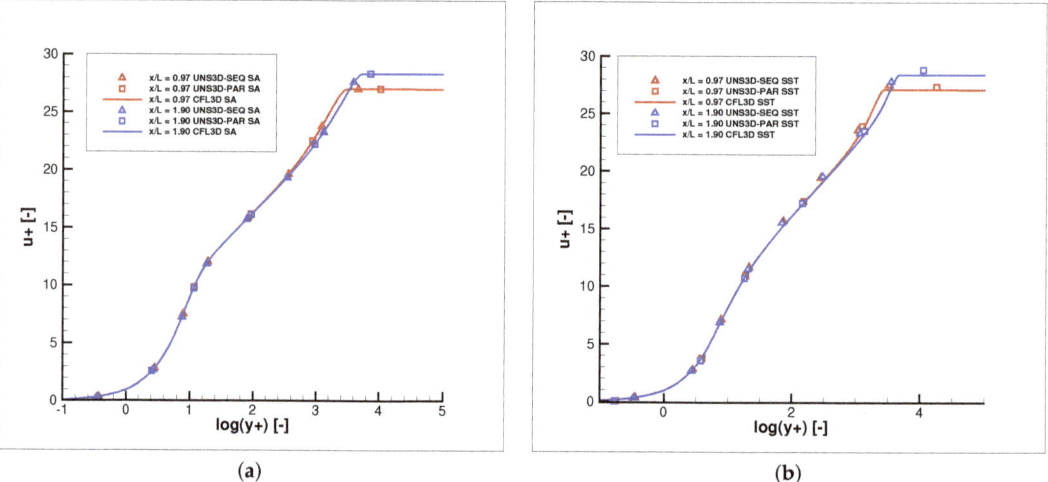

Figure 7. Law of the wake: (**a**) Spalart–Allmaras, and (**b**) Shear Stress Transport.

Figure 8 shows the skin friction coefficients predicted using the Spalart–Allmaras and Shear Stress Transport models. For the Spalart–Allmaras model, the skin friction coefficient was calculated using UNS3D-SEQ, UNS3D-PAR, and FUN3D codes. The agreement among the results generated by these three codes was excellent. For the Shear Stress Transport model, the skin friction coefficient was calculated using UNS3D-SEQ, UNS3D-PAR, and FUN3D codes. The skin friction coefficient predicted by the UNS3D-SEQ code was slightly smaller than the results predicted by the other two codes. Most likely, this difference is due to the fact that UNS3D-SEQ used a coarser grid than the other two codes, as shown in Table 3.

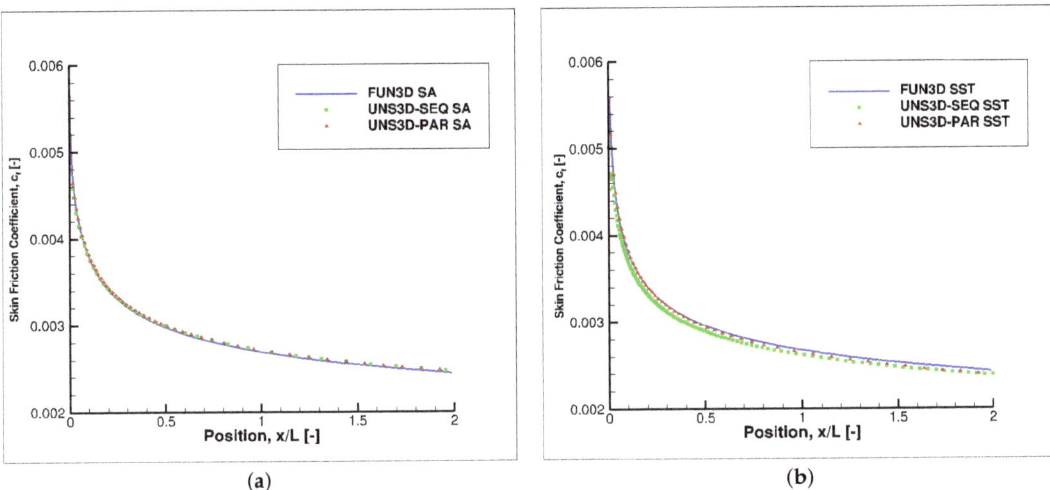

Figure 8. Skin friction coefficients for turbulent flat plate: (**a**) Spalart–Allmaras, and (**b**) Shear Stress Transport.

Figure 9 shows the predicted velocity profiles at two x/L locations: 0.97 and 1.90. The results generated with the UNS3D-SEQ, UNS3D-PAR, and CFL3D codes agree well, for both the Spalart-Allmaras and the Shear Stress Transport models.

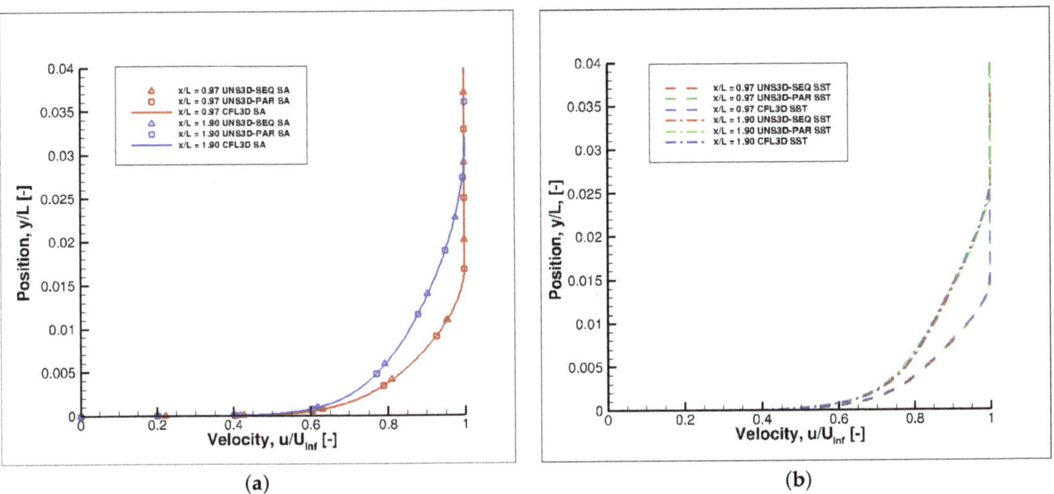

Figure 9. Velocity profiles on turbulent flat plate: (**a**) Spalart–Allmaras, and (**b**) Shear Stress Transport.

4.2. Two-Dimensional Airfoil Near-Wake

This test case is based off Model A airfoil, a convential airfoil that was experimentally investigated by Nakayama [22]. Model A airfoil is a 10%-thick conventional airfoil with a 61 cm chord. The angle of attack was 0 deg, the wind tunnel speed was 30 m/s, and the freestream turbulence level was 0.02%. The boundaries of the computational domain, shown in Figure 10, were 20 chords away from the airfoil. A detail of the mesh near the airfoil is shown in Figure 11. The input flow parameters used in the numerical simulation are listed in Table 4.

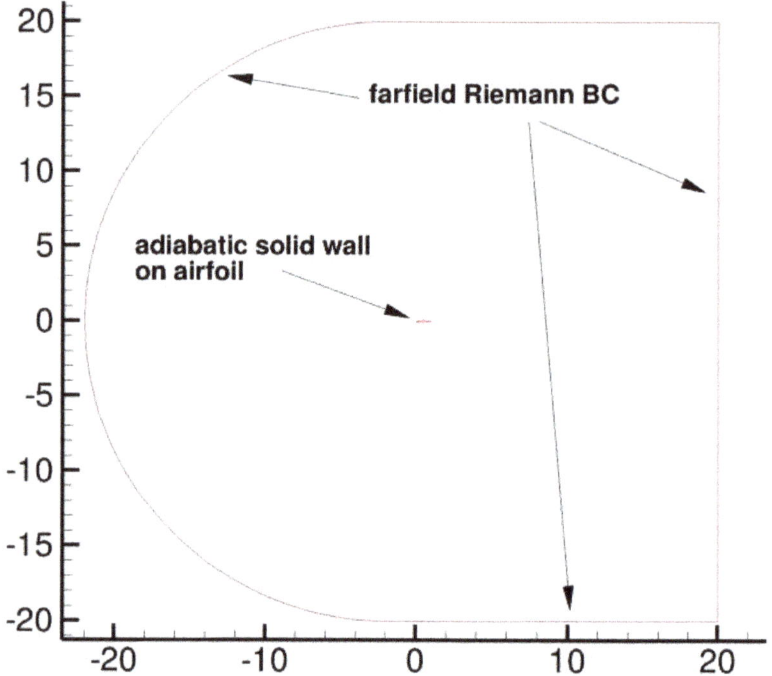

Figure 10. Airfoil boundary conditions [20].

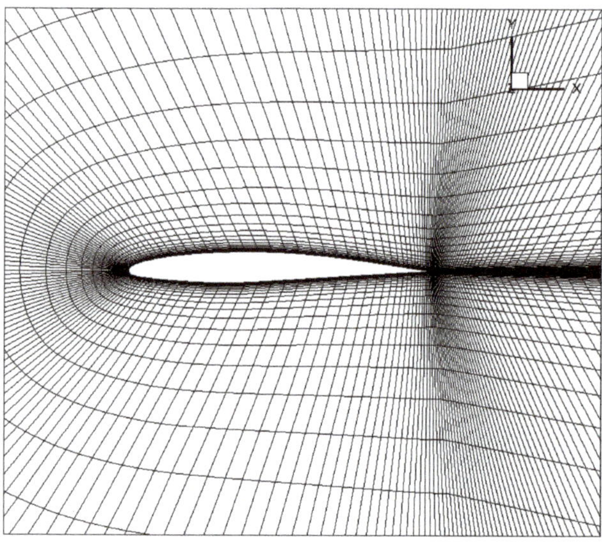

Figure 11. Detail of airfoil mesh.

Table 4. Input parameters for Nakayama airfoil.

p_{ref} [Pa]	T_{ref} [K]	M_{inlet} [-]	**Re** [-]	μ_{ref} [Pa s]	$\dfrac{\tilde{\nu}}{\nu}$ [-]	$\omega_\infty \dfrac{\nu_+}{c_\infty^2}$ [-]	$\kappa_\infty \dfrac{1}{c_\infty^2}$ [-]
62,000	300	0.088	1,200,000	1.838×10^{-5}	3	4×10^{-8}	4×10^{-10}

As with the turbulent flat plate case, the freestream values of κ and ω for UNS3D-PAR SST were different from the values used for the FUN3D simulation. Figure 12 shows the contour plots of the dimensionless turbulent viscosity μ_T / μ_∞ at the airfoil's trailing edge predicted using the Spalart–Allmaras model. There is good agreement between the results generated by the UNS3D-PAR and FUN3D codes. The turbulent viscosity predicted by the UNS3D-SEQ code, which used the coarse grid, was similar to the two codes; however, the lack of grid refinement at the trailing edge leads to a somewhat thicker wake.

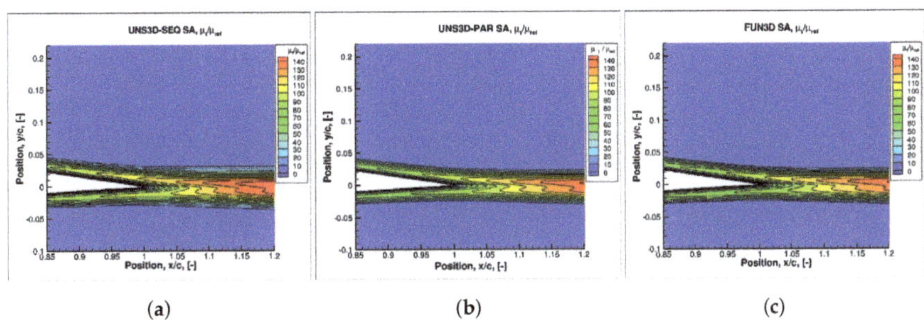

(a)   (b)   (c)

Figure 12. μ_t / μ_∞ contours on Nakayama's airfoil predicted with Spalart–Allmaras model: (a) UNS3D-SEQ, (b) UNS3D-PAR, and (c) FUN3D.

Figures 13 and 14 show the contour plots of the turbulent kinetic energy, κ, and the specific dissipation rate, ω, predicted using the Shear Stress Transport model. The agreement among the results generated by the three codes is good. While there are few differences on the turbulent kinetic energy contour plots, the specific dissipation rates are almost identical.

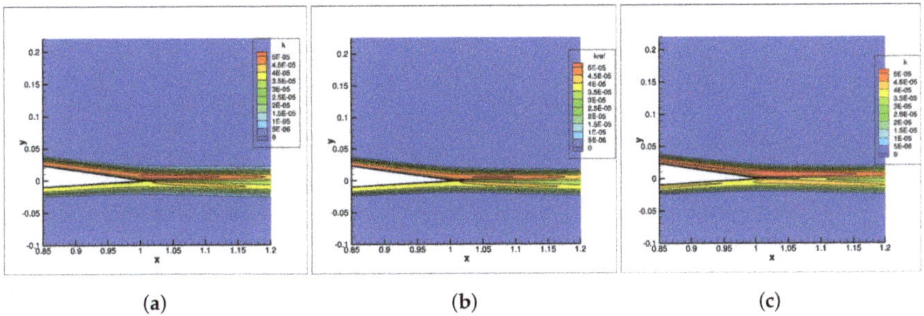

(a)   (b)   (c)

Figure 13. κ contours on Nakayama's airfoil predicted with Shear Stress Transport model: (a) UNS3D-SEQ, (b) UNS3D-PAR, and (c) FUN3D.

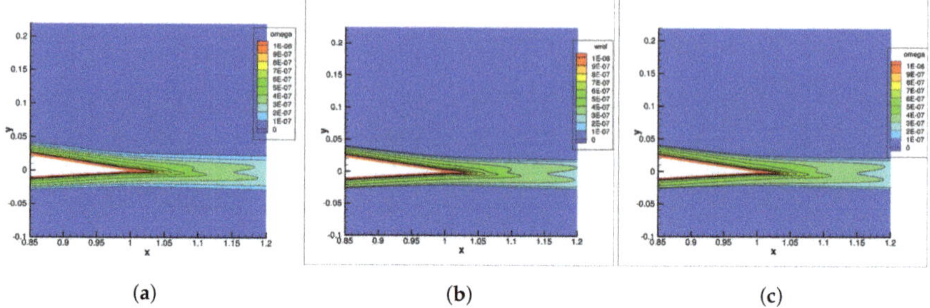

Figure 14. ω contours on Nakayama's airfoil predicted with Shear Stress Transport model: (**a**) UNS3D-SEQ, (**b**) UNS3D-PAR, and (**c**) FUN3D.

Figure 15 compares the measured velocity profiles against the predicted ones at seven locations in the airfoil's wake. These locations, measured from the airfoil leading edge, are: 1.01, 1.05, 1.20, 1.40, 1.80, 2.19, and 3.0 x/chord. The numerical predictions were done using the UNS3D-PAR code with the Spalart–Allmaras and the Shear Stress Transport models. The velocity profiles predicted using the Shear Stress Transport model matched the experimental data better than the Spalart–Allmaras model, although the Spalart–Allmaras model predicted the velocity profiles quite well.

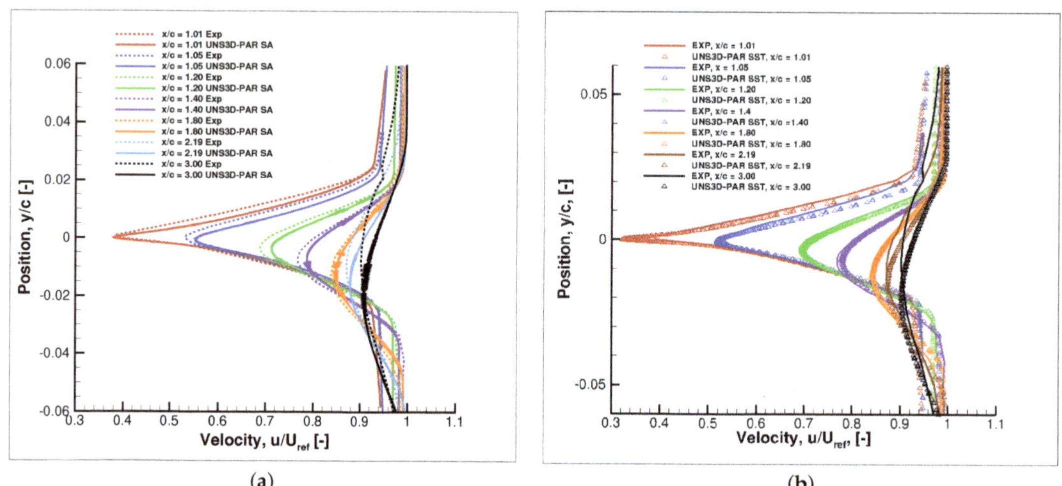

Figure 15. Velocity profiles on Nakayama's airfoil: (**a**) Spalart–Allmaras, and (**b**) Shear Stress Transport.

Figure 16 compares the measured turbulent shear stresses $-\overline{u'v'}$ against the predicted ones at the same seven locations in the wake. Both the Spalart–Allmaras and Shear Stress Transport models were used in the UNS3D-PAR code. The two turbulence models predicted the general shape of the shear stresses, although there were some differences between the predicted results and the experimental data.

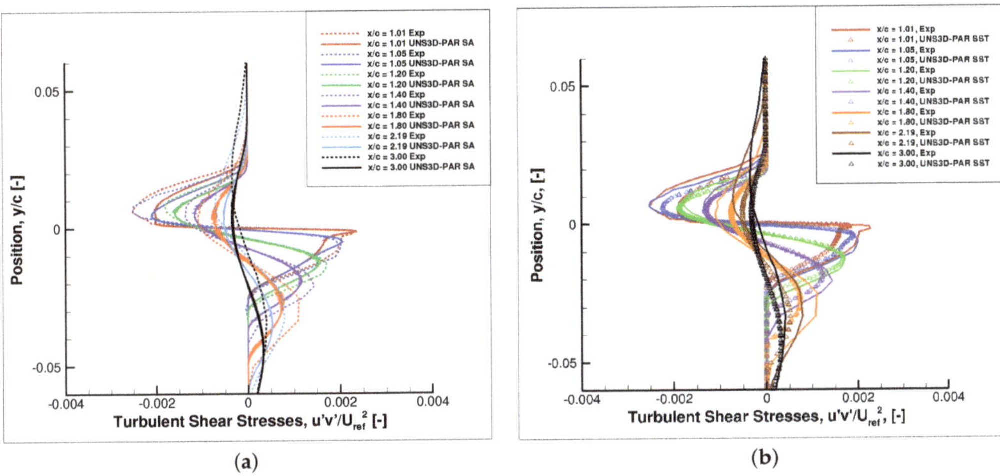

Figure 16. Dimensionless turbulent shear stress profiles on Nakayama's airfoil: (**a**) Spalart–Allmaras, and (**b**) Shear Stress Transport.

4.3. Two-Dimensional Backward-Facing Step

The backward-facing step is a case proposed by Driver and Seegmiller [23] to test how well the turbulence model simulates separated flows. The backward-facing step case is a challenging problem as turbulence models generally have difficulty modeling separated flows.

The test configuration had a 100 cm long × 15.1 cm wide × 10.1 cm high rectangular inlet duct followed by a backward-facing step of H = 1.27 cm. The geometry of the computational domain of the backward-facing step, shown in Figure 17, is defined as a function of the step height, H. The experiment took place at atmospheric total pressure and temperature. The freestream velocity was 44.2 m/s. The Reynolds number based on step height was Re_H = 36,000. The boundary conditions are also shown in Figure 17. A detail of the mesh near the step is shown in Figure 18.

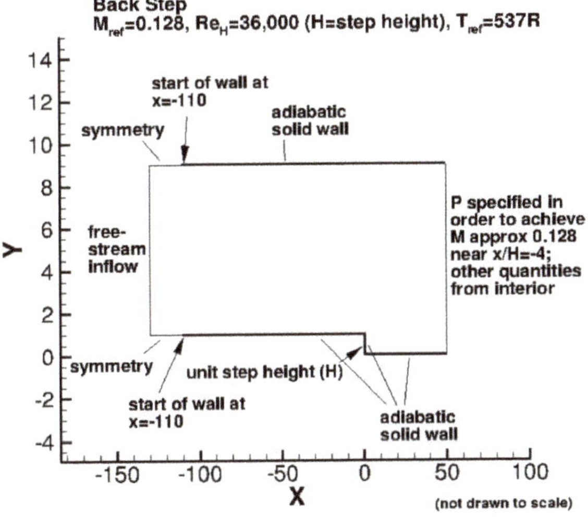

Figure 17. Backward-facing step geometry and boundary conditions [20].

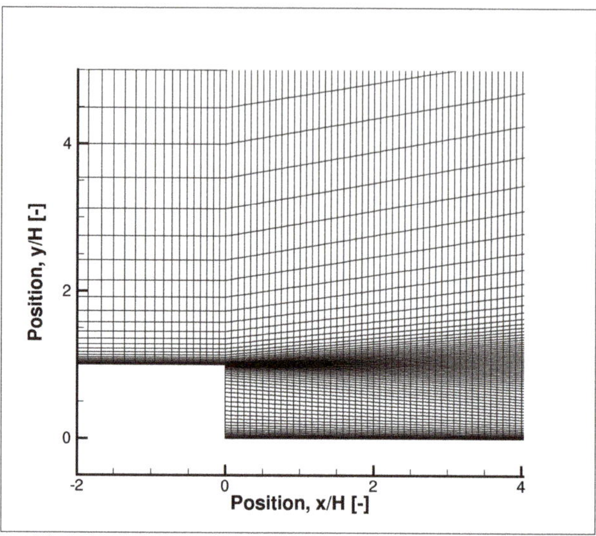

Figure 18. Detail of backward-facing step mesh.

Two grids were used for this test case: (1) the second coarsest and (2) the finest grid provided by the NASA Turbulence Modeling [20]. The results generated with the UNS3D-SEQ and UNS3D-PAR codes were compared against those generated with the NASA FUN3D and CFL3D codes which used the finest grid. The sets of results presented herein include velocity profiles, turbulent shear stresses, skin friction coefficients, and reattachment points.

Figure 19 compares the measured and predicted velocity profiles at four locations downstream from the step: $x/H = 1, 4, 6$, and 10. The velocity profiles were calculated using the UNS3D-SEQ, UNS3D-PAR, and CFL3D codes with the Spalart–Allmaras and Shear Stress Transport models. There was a good agreement among the results of the three codes, for both turbulence models. The Shear Stress Transport model matched the experimental data better than the Spalart–Allmaras model near the step, at $x/H = 1$. Away from the step, at $x/H = 4, 6$, and 10, the Spalart–Allmaras model captured the velocity profiles better than the Shear Stress Transport model.

Figure 20 compares the measured and predicted friction coefficients. There is a relatively good agreement among the results produced by the UNS3D-SEQ, UNS3D-PAR, and CFL3D codes, using both Spalart–Allmaras and Shear Stress Transport models.

The experimental investigation found the reattachment point to be at 6.26 ± 0.01 [23]. The predicted location of the reattachment point varied depending on the turbulence model. The Spalart–Allmaras model predicted an earlier reattachment, while the Shear Stress Transport model predicted a delayed reattachment. The predicted locations of reattachment point are summarized in Table 5.

Table 5. Location of reattachment point x/H on backward-facing step.

	Spalart–Allmaras	Shear Stress Transport
CFL3D	6.07	6.35
FUN3D	6.10	6.50
UNS3D-SEQ	6.27	6.79
UNS3D-PAR	6.01	6.83

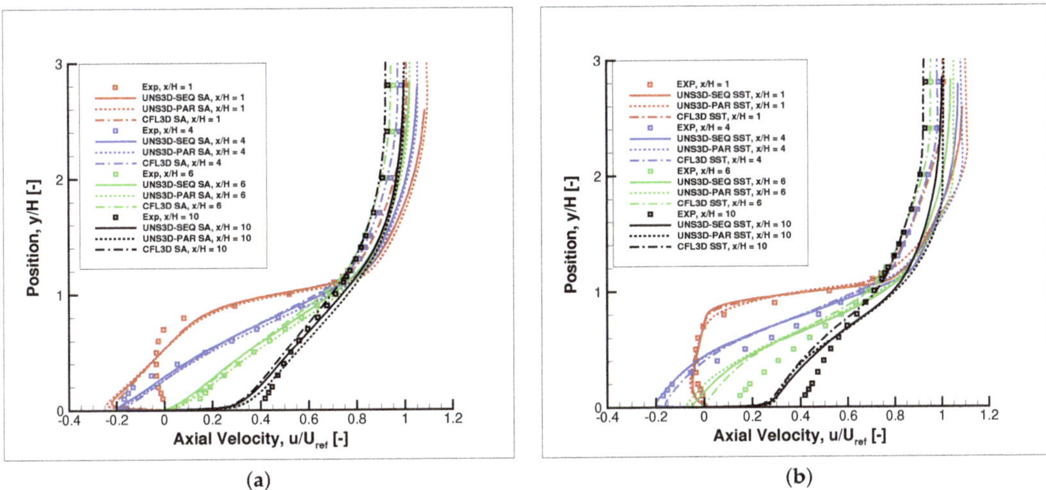

Figure 19. Velocity profiles on backward-facing step: (**a**) Spalart–Allmaras, and (**b**) Shear Stress Transport.

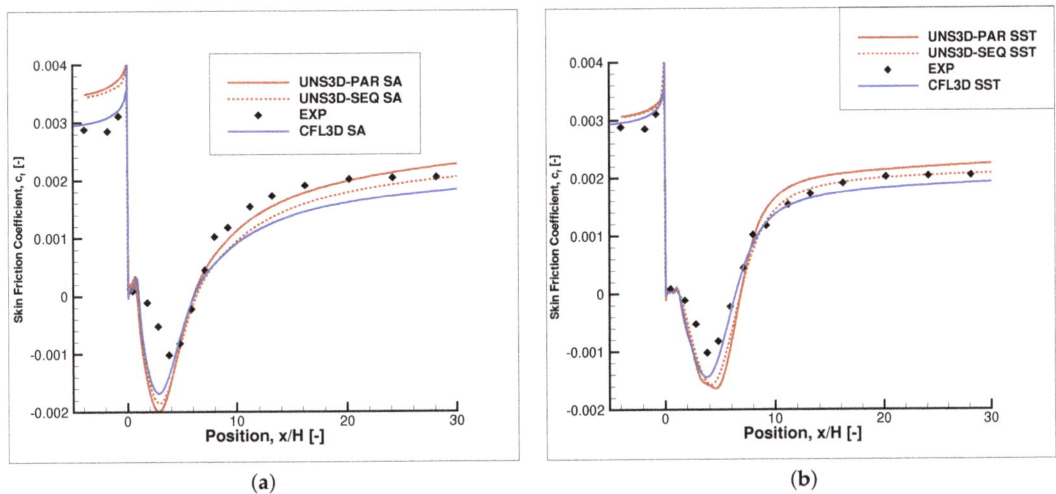

Figure 20. Coefficient of friction on backward-facing step: (**a**) Spalart–Allmaras, and (**b**) Shear Stress Transport.

Figure 21 compares the measured and predicted turbulent shear stresses at four locations downstream from the step. The results predicted using the Shear Stress Transport model matched the turbulent shear stresses better than the Spalart–Allmaras model.

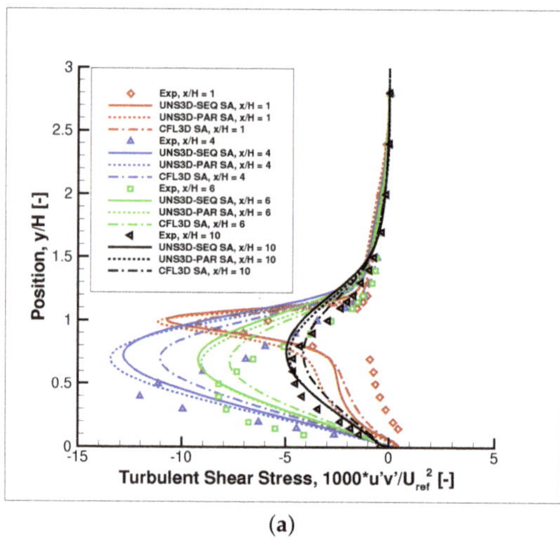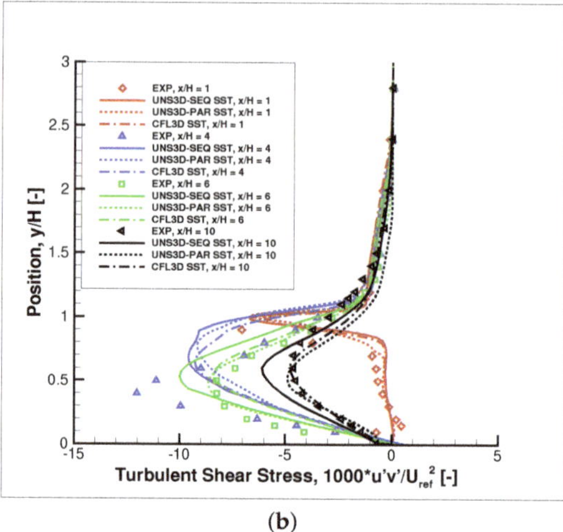

(a)   (b)

Figure 21. Dimensionless turbulent shear stress on backward-facing step: (**a**) Spalart–Allmaras, and (**b**) Shear Stress Transport.

4.4. Eleventh Standard Configuration

The Eleventh Standard Configuration contains two test cases from the experimental data obtained in the annular test cascade at EPF-Lausanne on a two-dimensional turbine cascade: (1) a subsonic, attached flow case, and (2) a transonic, separated flow case [24]. Herein, only the subsonic, attached flow case with stationary airfoils was simulated.

The airfoil chord is 77.8 mm, the pitch is 56.55 mm, and the stagger angle is $-40.85°$. The inflow angle is $15.2°$. The input flow conditions are listed in Table 6.

Table 6. Input parameters for eleventh standard configuration, case 100.

p_{tot}	T_{tot}	M_{inlet}	Re_L	μ_{ref}	$\frac{\tilde{v}}{v}$	$\omega_\infty \frac{v_L}{c_\infty^2}$	$\kappa_\infty \frac{1}{c_\infty^2}$
[Pa]	[K]	[-]	[-]	[Pa s]	[-]	[-]	[-]
124,600	330	0.31	446,000	1.846×10^{-5}	3	4×10^{-8}	4×10^{-10}

Figure 22 shows the computational mesh for two adjacent airfoils. The flow, however, was calculated over a single passage by using periodic boundary conditions to simulate the rest of the cascade. A multiblock O4H-grid padded with inlet and outlet H-grids was used to discretize the computational domain. The number of grid nodes varied between 39 k on the coarse mesh to 182 k on the fine mesh. Figure 22 shows the coarse grid. In the $x - y$ plane, the fine mesh used 401×101 nodes on the O-grid, 62×49 nodes on the block prior to the leading edge, 82×29 nodes on the block downstream from the trailing edge, 305×25 above and below the airfoil, 160×97 at inlet, and 200×77 at outlet. One cell was used the spanwise direction, as the flow was modeled as two-dimensional and the flow solver was three-dimensional.

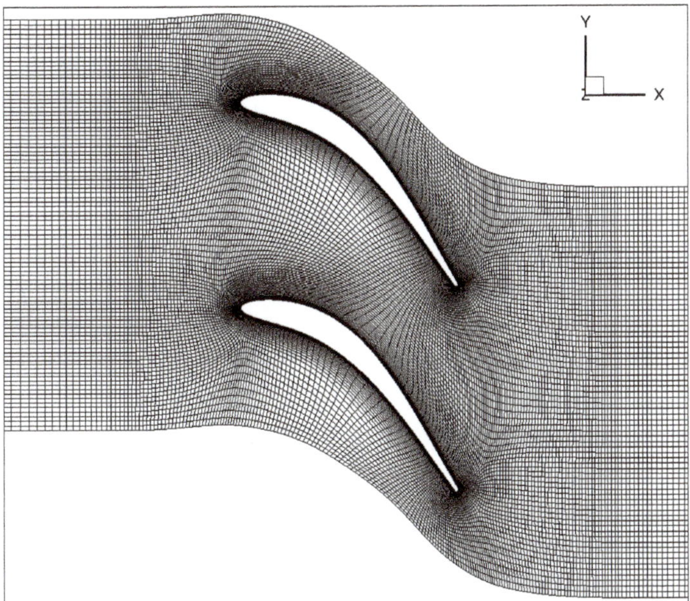

Figure 22. Eleventh standard configuration computational grid.

Figure 23 shows a comparison of the measured and predicted Mach number over the airfoil. The UNS3D-PAR code was used with the Spalart–Allmaras and Shear Stress Transport turbulence models. The results generated by the coarse and fine meshes were almost identical. Furthermore, the results generated by the two turbulence models were almost identical. On the pressure side, the UNS3D-PAR code results matched the experimental data well. Over the last one-quarter of chord, the predicted flow, however, had a higher velocity than the measured one. On the suction side, the predicted flow had a higher velocity than the measured one over 60% of the airfoil. The UNS3D-PAR code results matched the experimental data better than the numerical results reported in [24].

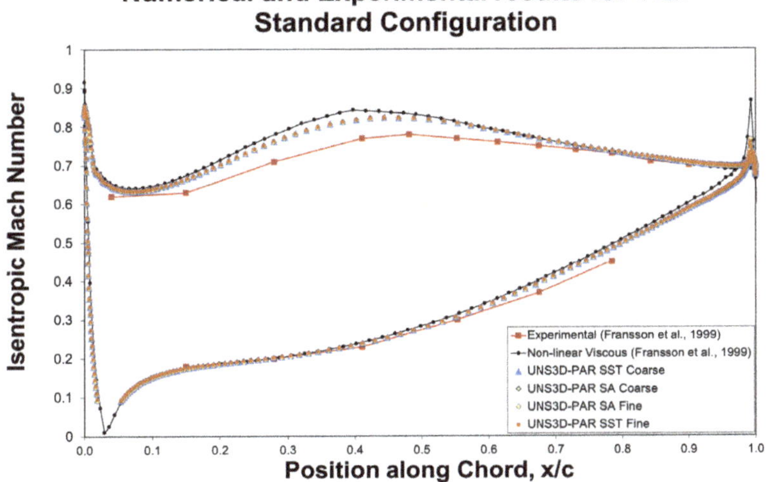

Figure 23. Measured and predicted Mach number on the eleventh standard configuration.

5. Conclusions

Four test cases were used to validate the implementation of the Shear Stress Transport and Spalart–Allmaras turbulence models in the UNS3D code. Two versions of the UNS3D code were considered: a sequential code, UNS3D-SEQ, and its parallel version, UNS3D-PAR. The validation of the implementation was done using experimental data and results generated by two NASA codes: FUN3D and CFL3D. The four cases considered were (i) a flat plate case, (ii) an airfoil near-wake, (iii) a backward facing step, and (iv) a turbine cascade known as the eleventh standard configuration. In all cases, the residuals of the solutions were less than 10^{-11}. The numerical results generated by the UNS3D codes with the two turbulence models compared well with the experimental data and the results of NASA codes FUN3D and CFL3D. The two turbulence models produced similar results. Compared to the experimental data, the Spalart-Allmaras turbulence model did not predict the turbulent fluctuations and skin friction coefficient as well as the Shear Stress Transport turbulence model. The Spalart–Allmaras model predicted the velocity profiles better than the Shear Stress transport model as the flow moved away from the backward-facing step. Overall, the results generated with the Shear Stress Transport model were closer to the experimental data, the only exception being the prediction of the reattachment point on the backward-facing step, where the Spalart–Allmaras model outperformed the Shear Stress Transport model. The computational time of the Shear Stress Transport model solution exceeded that of the Spalart–Allmaras model by 4% to 38%. In addition, the solution generated by the Shear Stress Transport model was more sensitive to the far-field boundary conditions than that of the Spalart–Allmaras model.

Author Contributions: Conceptualization: M.D.P. and P.G.A.C.; methodology: M.D.P. and P.G.A.C.; software: M.D.P. and P.G.A.C.; validation: M.D.P.; formal analysis: M.D.P.; investigation: M.D.P. and P.G.A.C.; resources: P.G.A.C.; data curation: M.D.P.; writing—original draft preparation: M.D.P.; writing—review and editing: M.D.P. and P.G.A.C.; visualization: M.D.P.; supervision: P.G.A.C.; project administration: P.G.A.C.; funding acquisition: P.G.A.C. All authors have read and agreed to the published version of the manuscript.

Funding: This work is supported by the Turbomachinery Research Consortium.

Acknowledgments: The authors would like to thank the High-Performance Research Computing center at Texas A&M University for providing the computational resources that were required to complete the work presented in this paper.

References

1. Jones, W.P.; Launder, B.E. The prediction of laminarization with a two-equation model of turbulence. *Int. J. Heat Mass Transf.* **1972**, *15*, 301–314. [CrossRef]
2. Johnson, D.A.; King, L.S. A Mathematically Simple Turbulence Closure Model for Attached and Separated Turbulent Boundary Layers. *AIAA J.* **1985**, *23*, 1684–1692. [CrossRef]
3. Wilcox, D.C. Reassessment of the Scale-Determining Equation for Advanced Turbulence Models. *AIAA J.* **1988**, *26*, 1299–1310. [CrossRef]
4. Menter, F.R. Two-Equation Eddy-Viscosity Turbulence Models for Engineering Applications. *AIAA J.* **1994**, *32*, 1598–1605. [CrossRef]
5. Spalart, P.R.; Allmaras, S.R. A One-Equation Turbulence Model for Aerodynamic Flows. In Proceedings of the 30th Aerospace Sciences Meeting & Exhibit, Reno, NV, USA, 6–9 January 1992; Number AIAA Paper 89-0366.
6. Wilcox, D.C. *Turbulence Modeling for CFD*, 2nd ed.; DCW Industries: La Cañada, CA, USA, 2000.
7. Allmaras, S.; Johnson, F.; Spalart, P. Modifications and clarifications for the implementation of the Spalart-Allmaras turbulence model. In Proceedings of the Seventh International Conference on Computational Fluid Dynamics (ICCFD7), Big Island, HI, USA, 9–13 July 2012; pp. 1–11.
8. Polewski, M.D. Time-Linearization of the Spalart-Allmaras Turbulence Model. Master's Thesis, Texas A&M University, College Station, TX, USA, 2020.
9. Kim, K.S. Three-Dimensional Hybrid Grid Generator and Unstructured Flow Solver for Compressors and Turbines. Ph.D. Thesis, Texas A&M University, College Station, TX, USA, 2003.
10. Roe, P.L. Approximate Riemann Solvers, Parameter Vectors, and Difference Schemes. *J. Comput. Phys.* **1981**, *43*, 357–372. [CrossRef]

11. Harten, A. Self Adjusting Grid Methods for One Dimensional Hyperbolic Conservation Laws. *J. Comput. Phys.* **1983**, *50*, 235–269. [CrossRef]
12. Barth, T.J.; Jesperson, D.C. The Design and Application of Upwind Schemes on Unstructured Meshes. In Proceedings of the 27th Aerospace Sciences Meeting, Reno, NV, USA, 9–12 January 1989; AIAA Paper 89-0366.
13. Venkatakrishnan, V. Convergence to Steady-State Solutions of the Euler Equations on Unstructured Grids with Limiters. *J. Comput. Phys.* **1995**, *118*, 120–130. [CrossRef]
14. Carpenter, F.L. Practical Aspects of Computational Fluid Dynamics for Turbomachinery. Ph.D. Thesis, Texas A&M University, College Station, TX, USA, 2016.
15. Dervieux, A. Steady Euler Simulations Using Unstructured Meshes. In *Computational Fluid Dynamics;* VKI Lecture Series 1985-04; Essers, J.A., Ed.; von Karman Institute for Fluid Dynamics: Rhode Saint Genese, Belgium, 1985; Volume 1.
16. Gargoloff, J.I. A Numerical Method for Fully Nonlinear Aeroelastic Analysis. Ph.D. Thesis, Texas A&M University, College Station, TX, USA, 2007.
17. Anderson, J.D. *Computational Fluid Dynamics: The Basics with Applications*; McGraw-Hill: New York, NY, USA, 1995.
18. Carpenter, F.L.; Cizmas, P.G.A. UNS3D Simulations for the Third Sonic Boom Prediction Workshop Part II: C608 Low-Boom Flight Demonstrator. In Proceedings of the AIAA Aviation Forum, Reston, VA, USA, 15–19 June 2020. [CrossRef]
19. Blazek, J. *Computational Fluid Dynamics: Principles and Applications*, 1st ed.; Elsevier: Amsterdam, The Netherlands, 2001.
20. NASA. NASA Turbulence Modeling Resource. 2019. Available online: https://turbmodels.larc.nasa.gov (accessed on 13 March 2019)
21. Coles, D. The law of the wake in the turbulent boundary layer. *J. Fluid Mech.* **1956**, *1*, 191–226. [CrossRef]
22. Nakayama, A. Characteristics of the Flow around Conventional and Supercritical Airfoils. *J. Fluid Mech.* **1985**, *160*, 155–179. [CrossRef]
23. Driver, D.M.; Seegmiller, H.L. Features of a reattaching turbulent shear layer in divergent channel flow. *AIAA J.* **1985**, *23*, 163–171. [CrossRef]
24. Fransson, T.H.; Jöcker, M.M.; Bölcs, A.A.; Ott, P.P. Viscous and Inviscid Linear/Nonlinear Calculations Versus Quasi-Three-Dimensional Experimental Cascade Data for a New Aeroelastic Turbine Standard Configuration. *J. Turbomach.* **1999**, *121*, 717–725. [CrossRef]

applied sciences

Article

Specific Modeling Issues on an Adaptive Winglet Skeleton

Salvatore Ameduri [1], Ignazio Dimino [1,*], Antonio Concilio [1], Umberto Mercurio [2] and Lorenzo Pellone [2]

[1] Department of Adaptive Structures, The Italian Aerospace Research Centre (CIRA), Via Maiorise, 81043 Capua, Italy; s.ameduri@cira.it (S.A.); a.concilio@cira.it (A.C.)
[2] Department of Aeronautical Technologies Integration, The Italian Aerospace Research Centre (CIRA), Via Maiorise, 81043 Capua, Italy; u.mercurio@cira.it (U.M.); l.pellone@cira.it (L.P.)
* Correspondence: I.Dimino@cira.it

Abstract: Morphing aeronautical systems may be used for a number of aims, ranging from improving performance in specific flight conditions, to keeping the optimal efficiency over a certain parameters domain instead of confining it to a single point, extending the flight envelope, and so on. An almost trivial statement is that traditional skeleton architectures cannot be held as a structure modified from being rigid to deformable. That passage is not simple, as a structure that is able to be modified shall be designed and constructed to face those new requirements. What is not marginal, is that the new configurations can lead to some peculiar problems for both the morphing and the standard, supporting, elements. In their own nature, in fact, adaptive systems are designed to contain all the parts within the original geometry, without any "external adjoint", such as nacelles or others. Stress and strain distribution may vary a lot with respect to usual structures and some particular modifications are required. Sometimes, it happens that the structural behavior does not match with the common experience and some specific adjustment shall be done to overcome the problem. What is reported in this paper is a study concerning the adaptation of the structural architecture, used to host a winglet morphing system, to make it accomplish the original requirements, i.e., allow the deformation values to be under the safety threshold. When facing that problem, an uncommon behavior of the finite element (FE) solver has been met: the safety factors appear to be tremendously dependent on the mesh size, so as to raise serious questions about the actual expected value, relevant for the most severe load conditions. On the other side, such singularities are more and more confined into single points (or single lines), as the mesh refines, so to evidence somehow the numerical effect behind those results. On the other side, standard engineering local methods to reduce the abovementioned strain peaks seem to work very well in re-distributing the stress and strain excesses to the whole system domain. The work does not intend to give an answer to the presented problem, being instead focused on describing its possible causes and its evident effects. Further work is necessary to detect the original source of such inconsistencies, and propose and test operative solutions. That will be the subject of the next steps of the ongoing research.

Keywords: adaptive structures; FE modeling; stress analysis; loads assessment

Citation: Ameduri, S.; Dimino, I.; Concilio, A.; Mercurio, U.; Pellone, L. Specific Modeling Issues on an Adaptive Winglet Skeleton. *Appl. Sci.* **2021**, *11*, 3565. https://doi.org/10.3390/app11083565

Academic Editor: Ruxandra Mihaela Botez

Received: 16 March 2021
Accepted: 12 April 2021
Published: 15 April 2021

Publisher's Note: MDPI stays neutral with regard to jurisdictional claims in published maps and institutional affiliations.

1. Introduction

The long experience gained by an extended Italian group on morphing, developed from more than 20 years of activity in international contexts, led to the ambitious aim of proposing a project concerning the design and realization of a complete morphing wing implementing different kinds of devices in its basic geometry. The occasion matured within the CS2 GRA frame, as Leonardo S.p.A. (Aircraft Division) issued a call for proposals on several topics, among which the abovementioned subjects [1]. CIRA, University of Napoli, and Politecnico di Milano took charge of designing and realizing four different adaptive prototypes to be installed on a single wing: a winglet, an adaptive trailing edge, and an adaptive leading edge plus an adaptive wingtip, respectively. The project did rely at a

61

wide extent on the further development of numerical capabilities particularly suited for morphing system design, on the wake of the developed expertise.

Kinematic systems differ from compliant morphing systems, as their basic architecture is almost an extension of standard devices already placed on any kind of aircraft, such as flap, slats, and so on, while the latter tries to modify the inner structure of the reference craft to make it suitably deformable [2–6]. In both cases, however, the need of assigning further macroscopic degrees of freedom to the target structure, imposes the creation of high-nodal density models [7]. They have to consider the deployment of a more complex and articulated structural outline, an immediate consequence of the additional mobility and the increased number of parts.

These kinds of models have been reported in the literature to be characterized by a certain behavior with respect to further increase of their density [8–11]. As the number of nodes and elements increase, or conversely, the mesh step decreases, certain singularities are shown that turn on the debate among the researchers about the fact they are actual index of a physical behavior or instead are simply linked to some mathematical issue [12]. The fact is, revealing the point deformation in a very limited geometry domain is a true challenge considering that standard sensors, even when miniaturized, only produce information along a certain area and do not have the capability of providing point information. Of course, this can be limitedly true for displacements, while they may have a dramatic impact on strain detection [13,14].

In light of these considerations, the paper herein presented reports some significant experiences that the authors believe to be interesting to discuss, concerning specifically the increasingly higher modal density mesh effects on the structural response. This goal had an even more significance, because of the possibility of attaining a wide database of experimental output, as the target element did undergo an extensive test campaign. As introduced, lab acquisitions were not expected to allow a deeper correlation among experimental and numerical data, but it was considered that could at least make a base of reference, even for further activities.

As part of the Clean Sky 2 program, CIRA is the leader of the AIRGREEN2 project in the REG-IADP platform whose leader is Leonardo S.p.A. (Aircraft Division). The AIR-GREEN2 project began in November 2015 and will end in 2023. The main purpose of the project is the development and validation of technologies suitable for the wing of the future regional aircraft of LDO and related demonstrators up to, for some of them, validation by experimentation in flight. The technological strands that characterize the AIRGREEN2 project are:

Innovative wing structure (design, manufacturing and testing): innovative composite design, repair, structural health monitoring, LRI (liquid resin infusion) processes and AFP (automatic fiber placement) [15,16].

Innovative aerodynamic design: extended laminar flow, 3D riblets, morphing winglet, morphing flap to stay in condition of maximum aerodynamic efficiency during the cruise, development of techniques for load control and alleviation (LCA system based on an innovative wing tip and morphing winglet) [17].

Adaptivity: high lift performance using an innovative leading edge (droop nose) compatible with laminarity requirements and coupled to Morphing Flap

In line with the main targets of AIRGREEN2, specific morphing concepts have been being under investigation [1]. They will contribute to enhance the performance of the A/C, to optimize the load distribution on the wing structure and to increase the lift generation.

Four different concepts have been conceived:

(1) An adaptive leading edge, able to increase its curvature, to generate the so called droop nose effect [18]. The smooth morphed geometry dramatically contributes to the lift generation capability at specific operational conditions (take-off and landing), while the absence of geometric discontinuities in clean configuration cooperates in keeping the laminar flow in cruise. The device is constituted by a compliant internal

structure, moved by a spanwise torque actuator. The deformation is then transmitted to the surrounding skin, a carbon fiber laminate.

(2) A multifunctional flap, relying upon an adequate number of degrees of freedom (DOFs) to change its curvature, through a finger-like kinematic architecture [19]. The role of the device is to cooperate with the adaptive leading edge to generate additional lift (change of the curvature along the entire chord in fully extracted configuration) and to enhance the aerodynamic efficiency and optimize the load distribution (deflection of the aft tabs in fully retracted configuration).

(3) An adaptive wing tip, a 1 DOF aerodynamic movable surface mounted at the tip of the wing [20]. Deflected by an electromechanical actuator, its role is to cooperate with the multifunctional flap in the load control and alleviation operations.

(4) An adaptive winglet, mounted alternatively to the just mentioned wing tip and constituted by a fixed structure equipped with independent movable surfaces [21,22]. The scope is to adaptively alter the load distribution along the wing span to increase the aerodynamic performance, from one side, and to alleviate the stress level, from the other side.

The novelty of aforementioned devices is different for each of them. Adaptive trailing edges are very popular in literature and different architectures and concepts may be found in bibliography. The main characteristics of the one herein reported, and originally developed by the University of Napoli "Federico II", stands in its capability of combining up to three modalities of morphing (classical flap-like modification, tab-like movement, and artificial chord-extension), while keeping low the number of parts and therefore its complexity. The adaptive winglet developed by CIRA is almost an innovation itself, since the previous experiences on that topic mainly concern the introduction of a tab-like system, mainly vibrating at high frequencies in order to try gust effect alleviation strategies (Airbus Innovation Work, 2015), or introducing hinged wing tips for the modification of the configuration in-flight. In this case, the system is instead conceived to perform both static and dynamic functionalities along all the aircraft mission. It should be reminded that this kind of system shows the intrinsic challenge of implementing large weights at the extremity of the wing, therefore modifying intimately the aeroelastic behavior of reference aircraft.

The adaptive leading edge is characterized by an internal compliant architecture that cooperates with the external structure in absorbing loads and at the same time drives the deformation, leading to the drooped nose configuration. The device jointly to the multifunctional flap action leads to a significant hyper lift generation, moving upward the lift curve of the wing and increasing the stall angle.

The innovative wing tip due to its location on the wing and to its specific configuration is particularly prone to implementation of LCA strategies of new conception.

1.1. General Overview of the Winglet System

Among the just mentioned four AIRGREEN 2 morphing devices which may be considered for similar purposes, this study focuses on specific modeling issues addressed during the advanced design of a finger-like mechanism-based morphing winglet developed by CIRA. Such an integrated morphing system, capable to reduce wing-bending moments and increase aircraft flight stability in response to changing flight conditions, has proved to provide significant aerodynamic benefits, estimated on the order of 2.5% LoD at high CL with respect to the optimal passive winglet counterpart [23].

The morphing winglet system consists of two "finger-like" mechanisms realizing two independent morphing tabs moving both upward and downward. Such deflections are driven by dedicated electromechanical actuators [24]. One of the major advantages of this architecture is the capability to move the individual surfaces either synchronously or independently to different angles. This ensures smoother morphing winglet aeroshapes and a more efficient distribution of the span-wise aerodynamics, as estimated by computational fluid dynamics (CFD) simulations (Figure 1).

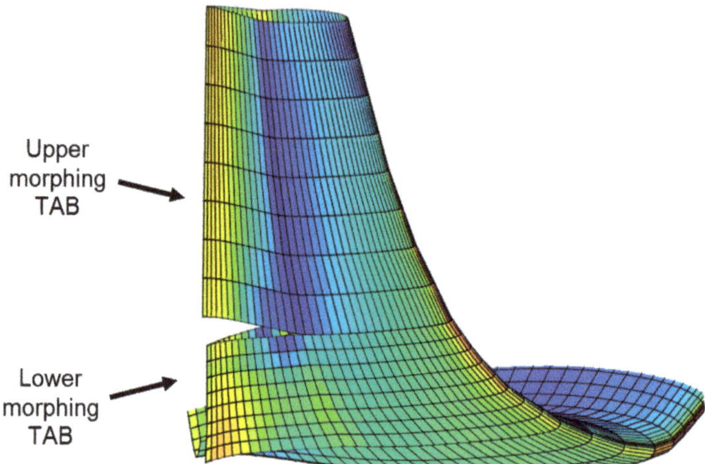

Figure 1. Morphing winglet concept with upper and lower control surfaces [24].

The structural layout of the morphing winglet consists of a passive and an active part [25]. The former is made of laminate skin panels and a torsion box consisting of spars and ribs. The latter incorporate two hinged mechanisms covered by a segmented skin. By means of linking rods, hinged on non-adjacent blocks, the morphing capability is enabled by the relative rotations of three adjacent blocks, which are free to rotate around the hinges on the camber line, thus physically turning the camber line into an articulated chain of consecutive segments. Each tab is, therefore, a single degree of freedom (SDOF) system; if rotation of any of the blocks is prevented, no change in shape can be obtained. On the contrary, if an actuator moves any of the blocks, all the other blocks follow the movement accordingly.

1.2. The Reference FE Model

For the purpose of this research, focus is given to the passive structure of the winglet, shown in Figure 2, sized to withstand the worst-case loading conditions that it is expected to experience in normal operation. The structural layout, consisting of upper and lower CFRP panels, internal CFRP spars and ribs, is overall modeled by 10,593 2D elements, defining a reference finite element (FE) model [23] considered in this study to investigate specific modeling issues.

Both the upper and lower skin panels consist of the same CFRP lay-up sequence represented by CQUAD4 and CTRIA3 elements in relation with laminate property entries (PCOMP) on the middle surface. All laminates are symmetric and are mainly composed of fabric layers. UD-plies in x direction are used for reinforcement. The material orientation is defined on the respective CQUAD4/CTRIA3 entry and is aligned along the span. All the elements used in the FE model are linear. The thickness of the laminates of the skins, spars and ribs is between 2.7 and 3.6 mm. The final weight of the complete structure is below the preliminary design target of 50 kg.

For the analysis of the laminates, a maximum/minimum strain criterion in 11- and 22-direction is used. Additionally, stress is also evaluated in 11-, 22- and 12-directions. The ply stresses and strains are read from the FEM results. The failure indexes for each ply were then calculated. The enveloped results were then elaborated to allow for computation of the minimum failure index.

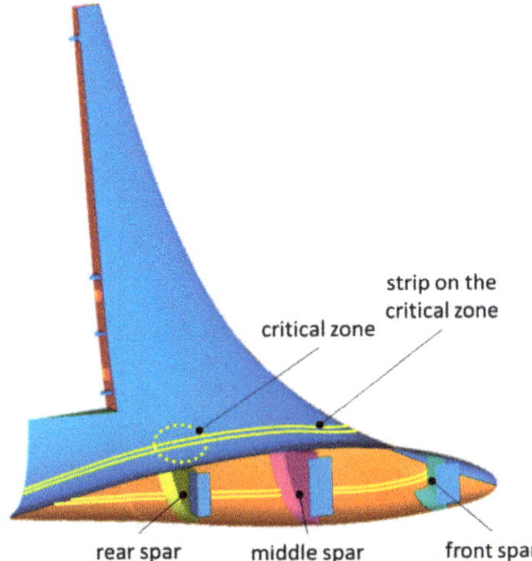

Figure 2. Finite element (FE) model of the winglet without the moveable surfaces [23].

1.3. Load System

Different load conditions have driven the design of the fixed part of the winglet, corresponding to the most severe solicitation at bending, torsion and shear, both in negative and positive directions. An initial trade-off has been carried out to estimate the impact of each load condition on the stress level generated into the main subparts of the structure. The radar plot, shown in Figure 3, illustrates a comparison of the stress produced by the just mentioned load cases, normalized with respect to the most severe case. As evident, the first load case, corresponding to the maximum bending in positive direction (inward deflection of the winglet) dramatically envelopes the other ones, followed by the load case 2, due to the bending too, but in the opposite direction. For this reason, only the 1st load case has been considered significant for the investigations illustrated in the next part of the work.

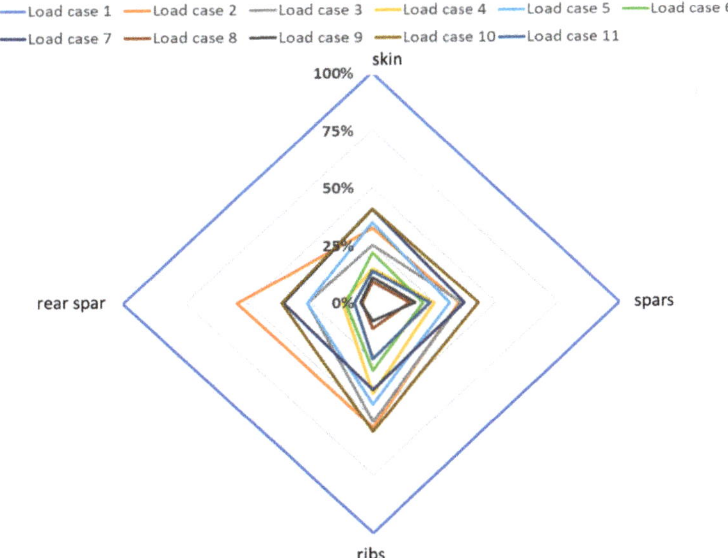

Figure 3. Load case comparison in terms of normalized stress level for the main subparts of the structure.

The skin subpart exhibited the highest level of stress: a peak has been noticed on the upper zone of the skin, close to the cap of the rear spar, leading to a stress concentration. It has been just this outcome to justify the investigation hereafter presented.

2. Criticalities and Approach

The just mentioned stress concentration has occurred on the zone highlighted in Figure 4, depicting the stress map for the aforementioned load case 1. This stress distribution has been assumed as the standard reference load (RL). The type of solicitation is in line with the bending action of the load case, that produces a compression of the skin in span direction and its bending around the spar cap along the chord. This, jointly to the dramatic variation of the local thickness (passing from the overlap of the skin laminate and the cap to the skin alone), can justify the stress concentration.

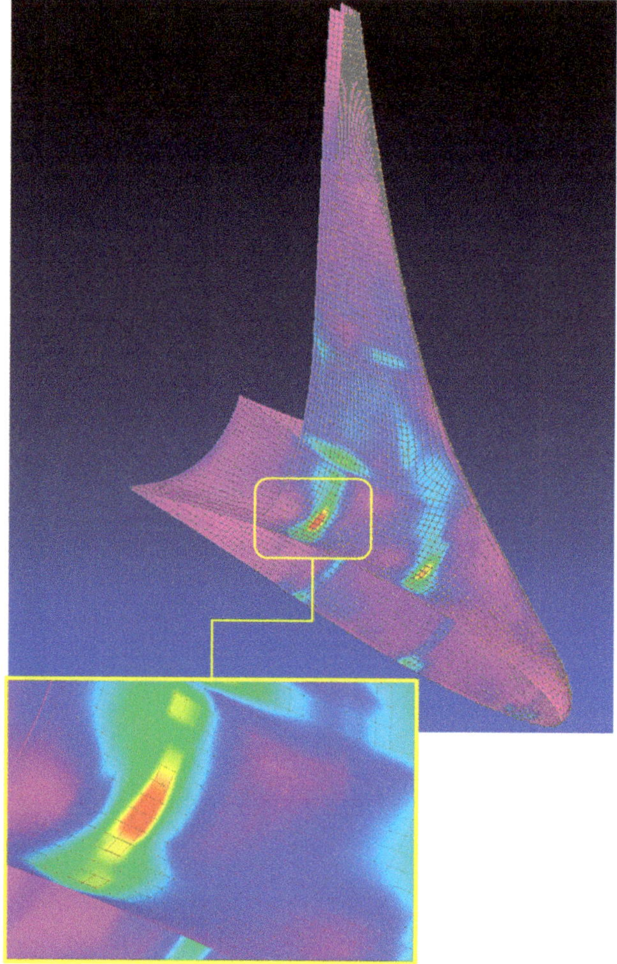

Figure 4. Stress concentration on the upper zone of the skin, close to the rear spar.

However, it is questionable whether the discretization could have a role on the estimate of the stress level. To answer this question a parametric study has been organized, considering progressive refinement levels of the mesh on the zone of interest. This skin region, as shown in Figure 5, is constituted by two consecutive lines of seven 19×19 mm plate elements, towards the span direction. It covers the skin part upstream the spar cap and corresponds to the critical zone highlighted in Figure 4.

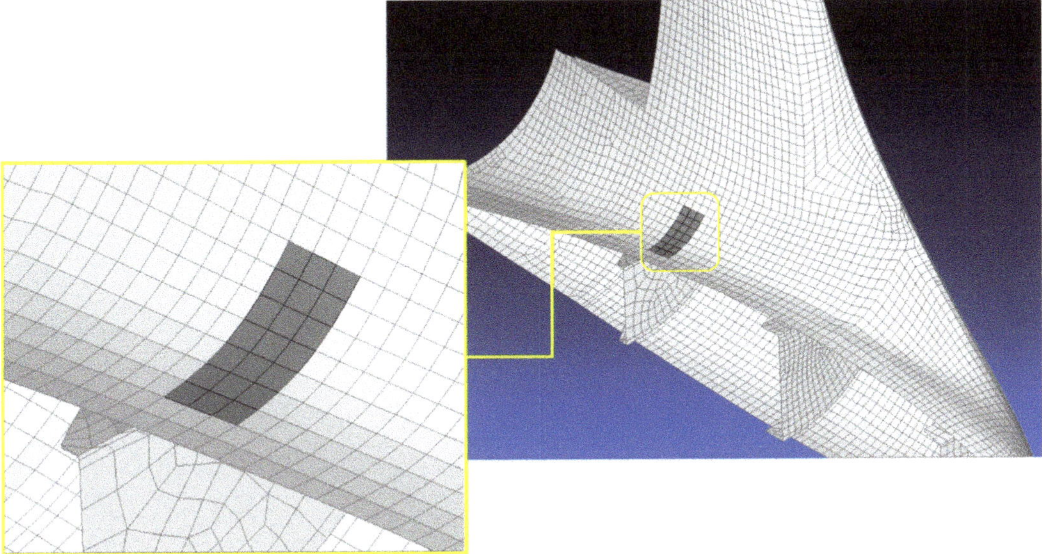

Figure 5. Zone of the model considered for the parametric refinement.

At first, the elements have been split along the chord, since the highest stress gradient occurs along this direction. Then a further split has been performed along the spar, obtaining again square elements. This process -split along the chord and then along the span- has been repeated for two times, coming to the end to square elements 1/4 sized with respect to the original ones. To guarantee an adequate continuity with respect to the surrounding mesh, the boundary elements have been suitably split into triangles, fitting the nodes on the edges.

The stress maps illustrated in Figure 6 show the effect of the level of refinement in terms of stress and distribution. The maximum normalized stress obtained has been summarized in Table 1 and plotted in Figure 7.

Table 1. Maximum normalized stress vs. mesh refinement level; normalization reference: maximum stress computed on the standard reference load (RL) distribution reported in Figure 4.

Case ID	Element Size (mm)	Normalised Stress (%)
reference	19×19	100.0
case a	19×9.5	107.3
case b	9.5×9.5	97.3
case c	9.5×4.75	130.9
case d	4.75×4.75	154.5

(*) The second dimension refers to the chord direction.

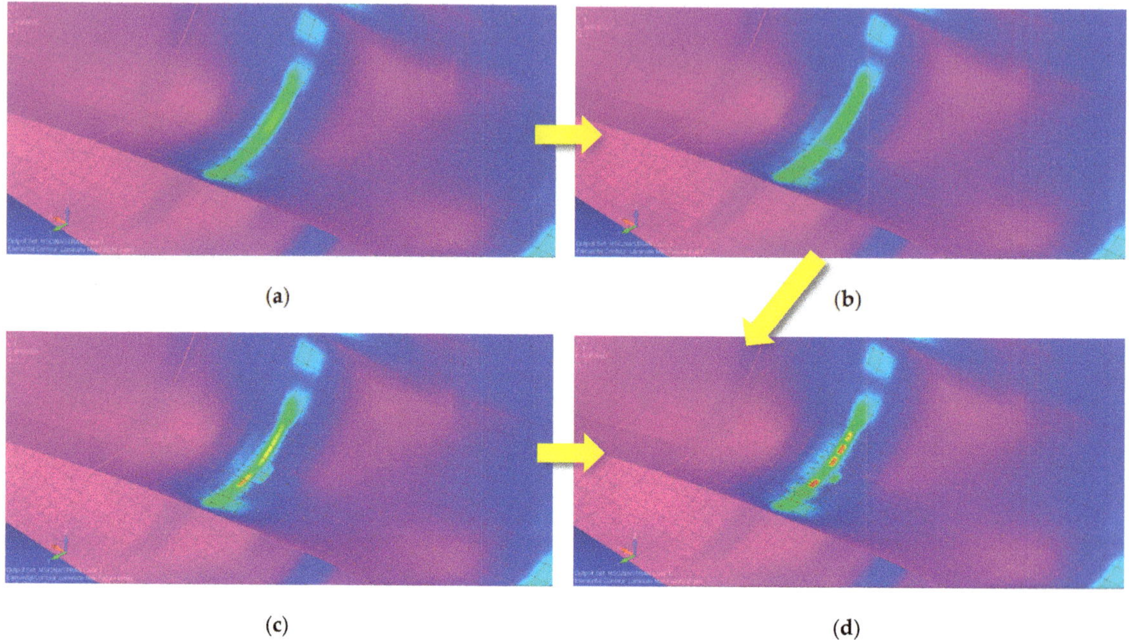

(a)

(b)

(c)

(d)

Figure 6. Normalized stress maps vs. mesh refinement level: 1st chordwise split (**a**), 1st spanwise split (**b**), 2nd chordwise split (**c**), 2nd spanwise split (**d**); normalization reference: maximum stress computed on the stress distribution reported in Figure 4 (RL).

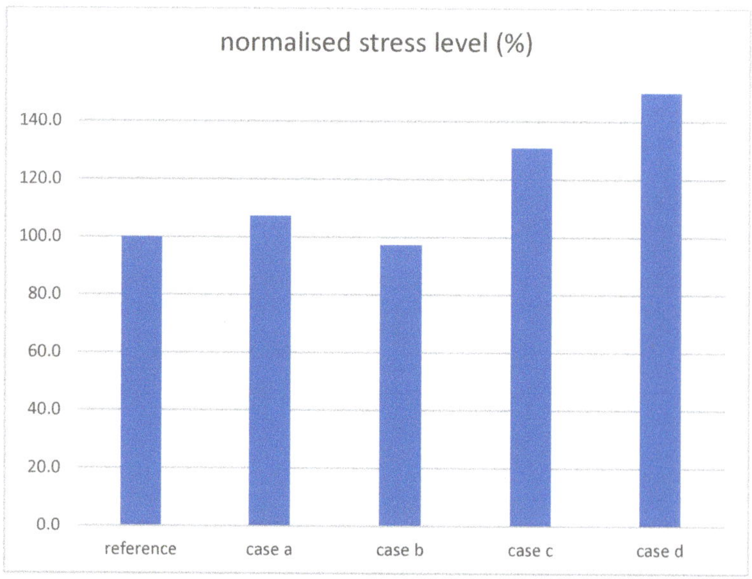

Figure 7. Normalized maximum failure index vs. mesh refinement level: 1st chordwise split (**a**), 1st spanwise split (**b**), 2nd chordwise split (**c**), 2nd spanwise split (**d**); normalization reference: maximum stress computed on the distribution reported in Figure 4 (RL).

The refinement process highlighted, in general, an increasing trend in terms of stress level. This can be explained considering two numerical aspects. The first one is represented by the strain definition as ratio (gradient) of the local displacement over the element size: as this latter term goes to zero over a certain threshold, the division badly handled by the solver. The second aspect is strictly related to the inversion of the structural matrix, whose conditioning level is dramatically affected by its size. This aspect, for the specific case under investigation, however, does not play a fundamental role, since the increase of DOFs due to the local refinement is really modest with respect to the original size of the structural matrix.

The stress maps of the different cases show a concentration on the central part of the investigated zone, with a uniform distribution on the transition between the central zone and the rest of the model. This in some way resizes any mesh boundary effect and focuses the attention on the central part, affected by the discontinuity of stiffness. Another interesting aspect is the impact of both the in-plane element dimension onto the stress. The cases in Figure 6b–d, in fact, show a linear trend punctuated by progressive halving of both the thickness gradient and the in-plane normal directions. This suggests that both the thickness variation and the element size play a role on the observed behavior.

The high level of stress, jointly to the uncertainty related to the just mentioned trend with the refinement of the mesh has led to the adoption of a design solution to mitigate the local accumulation of stress.

As already discussed, the investigated load case determines the inward bending of the winglet, with a consequent compression of the top skin along the span direction. A secondary effect is the bending of the skin also in chordwise direction, between the rear and the middle spars. Considering that the stiffness is higher in the spanwise direction for the overlap of the cap and of the skin laminate and that a sudden variation of rigidity occurs upstream the rear spar, solutions altering the chordwise stiffness look more appropriate. In line with this consideration, a "T" shaped beam bonded on the inner surface of the skin and linking the rear to the middle spar has been investigated. The layout considered is illustrated in Figure 8.

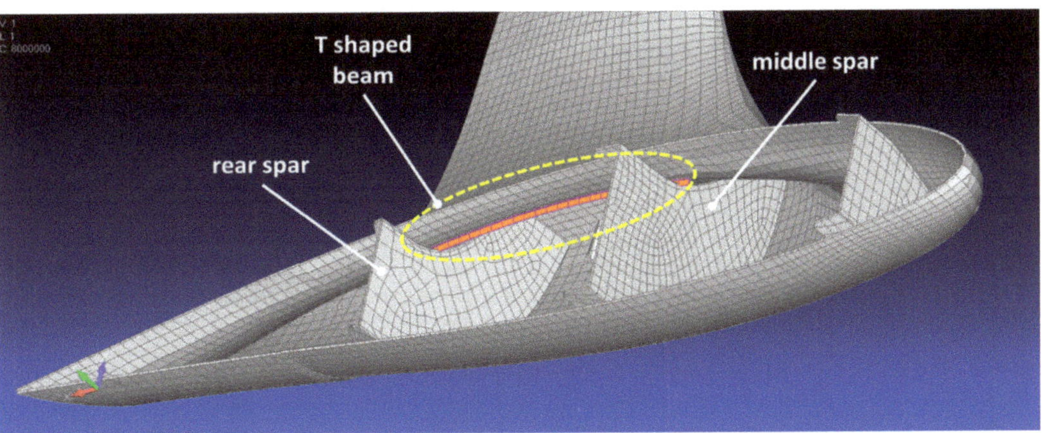

Figure 8. Layout of the chordwise beam to increase the rigidity on the critical zone.

The solution has proved to reduce of about 37% the stress level, thus definitely mitigating the local criticality. A comparison of the initial and final configuration is presented in Figure 9, that highlights a better situation not only on the rear spar but also on the middle spar.

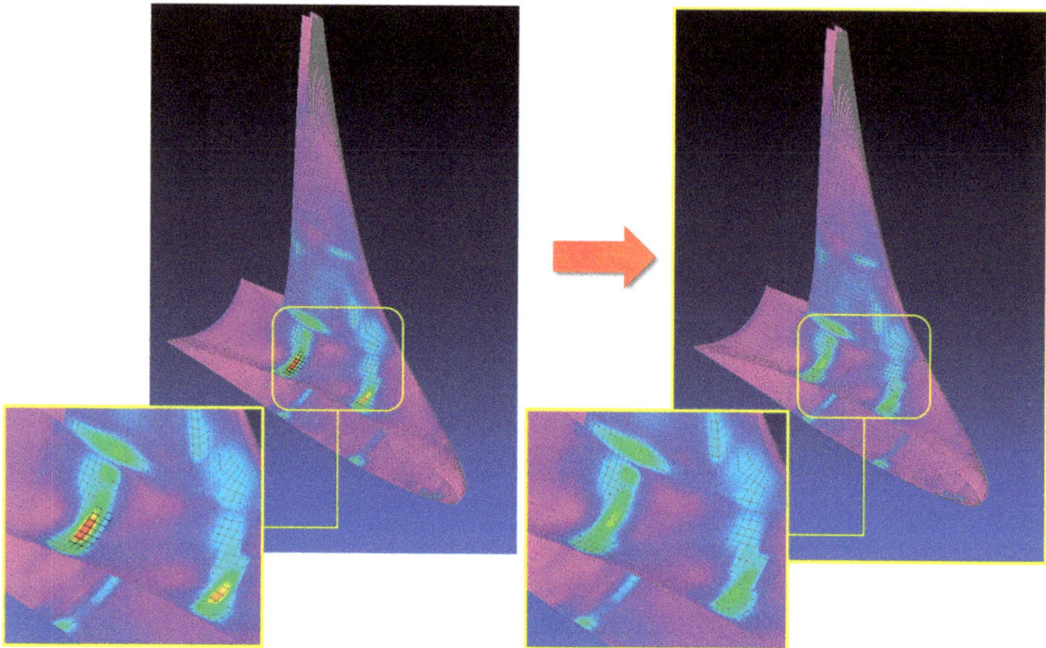

Figure 9. Normalized stress for the original configuration, RL, (**left**) and the one integrated with the additional beam (**right**); normalization reference: maximum stress computed on the distribution reported in Figure 4 (RL).

3. Impact, Discussion and Perspective Works

The just mentioned investigations in some way highlight two aspects that concur to make critical the zone of the skin explored: the sudden and dramatic variation of the local thickness, due to the sudden overlap of the spar-cap to the skin, and the level of discretization adopted to model the zone. The behavior shown by the model leads to severe consideration on its reliability and on the exploitation of the achieved results.

The reliability problem should be faced in two steps: firstly, confining the portion of the model that can be considered computationally robust and, secondly, defining a strategy of elaboration of the results on the critical zone. A model completely refined could help support the first step, in some way confirming the accuracy of the predictions on a certain portion of the winglet domain. A dedicated strategy, on the contrary, should be identified for the second step. Here, the just mentioned concurring aspects (mesh refinement level and thickness variation) should be accurately weighted, jointly to the type of connection between the skin and the spar cap (rigidly or elastically linked, or seen as a unique layup) and, most importantly, the way the load fluxes distribute against the cross-section areas. This last point seems to be the nearest to the physics and suggests a real concentration of stress in the zone; however, the spike due to the thickness discontinuity, further confirmed and emphasized by the sharper and sharper gradient of progressive mesh refinements, should be realistically interpreted and smoothed by means of a moving average process. The plot of Figure 10 illustrates the effect of a moving average three-element windowed, on the FI computed critical strip highlighted in Figure 2. A remarkable mitigation of about 55% on the rear spar has been estimated. In any case, experimental information coming from strain gage local measurements, discussed in the next section should drive or at least support the reliability process of the model.

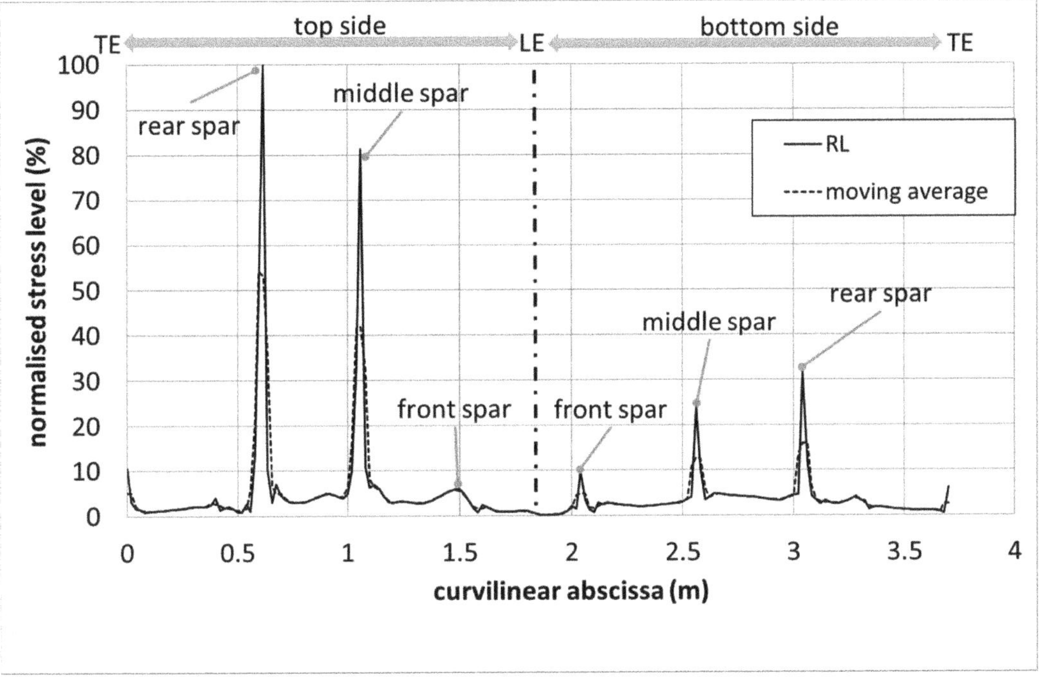

Figure 10. Normalized stress level vs. curvilinear abscissa of the winglet strip: original (solid) and after moving average (dashed); normalization reference: maximum stress computed on the distribution reported in Figure 4 (RL).

These issues suggested the investigators to move towards slightly different and more robust numerical configurations, that avoided the insurgence of those spikes.

4. Experimental Issues

The envisaged singularities were so defined mainly for their very limited, almost point extension. Even considering this characteristic can be hardly linked to a physical behavior, we cannot attribute this result to a pure mathematical issue. Conversely, it is very hard to reveal that behavior experimentally, and should it be not found, the doubt remains concerning its existence or not: is it the net sufficiently narrow to catch that small peak? This consideration leads to important consequences. Strain gauges are not appropriate to further investigate the phenomenon. They are significantly wider than the area of interest, and they have the bad characteristic integrated into the info over their area, so that a mean value is actually detected. Furthermore, their installation could somehow affect the measurements in itself, just for their finite area. A good alternative could be represented by the use of distributed fiber optics, that can exhibit very dense measurement arrays, stepped and extended a few mm. In that case, however, it should be considered that the measure could only be directional, while a 2D field could be acquired with an appropriate and well-defined deployment. Hand placement does not favor the accuracy of measures on very specific areas. Even more, the numerical results would suggest the need of an accurate investigation "by the single layers" of the composite structure. This further complexity could indeed be solved by embedding the thin fibers within several layers, at least in principle. Other methods of investigation could be also exploited, as for instance accurate laser map measurements, that has however the backlash of needing numerical derivation to attain strain information, with the further complexity that the structure is not slender and that the measured displacement is normal to the requested longitudinal strain.

The conclusion is that this kind of phenomena deserves dedicated investigations to be faced and finally solved. Available technology seems to provide some opportunities. Nevertheless, the test campaign shall be designed very accurately since the beginning to allow a certain level of success. An almost obvious, further consideration is that the reference object itself should be properly designed in order to match numerical and physical requirements, and be relevant on the investigation point of view.

5. Conclusions and Further Steps

The abovementioned results, and the surrounding conditions that they were obtained from, give a lot of interesting keys for further deepening of this kind of analysis.

First of all, as cited in the introduction, these kinds of meshes and the complex architecture is almost a constant for morphing systems; which means that this situation is doomed to be recurrent for this kind of modeling, irrespectively of the adopted configuration. Therefore, it is worthy of study and should be understood in detail.

The obvious consequence is that these kinds of outcomes will repeat more or less equally for the different architectures, so that a recursive impact on design and verification times is expected—then propagating to the development times.

The reported issue is not dramatic in itself, even if it should reveal to be coherent with the physical behavior, as it was shown that minimal intervention on the structural architecture may report the strain excess to usual and controllable values. Additionally, the effectiveness of that solution itself—directly affecting the local stiffness and in some way mitigating the discontinuity of rigidity—supports the hypothesis of a stress concentration due to the sudden variation of thickness. Indeed, the thickness variation has a relevant impact upon the stiffness matrix, due to the cubic dependence.

For the statement above, it could be argued that this issue could be eliminated by a simple implementation of minor adjustments, affecting the layout and the weight, minimally. The easy answer is that once the bug has been pointed out, it is safe to adequately investigate the problem, to avoid further, dramatic consequences in successive design processes.

Even if these outcomes should prove to be far from design reality, it could be speculated that they are signals of some specific structural modality behavior, which could emerge as some damage or discontinuity occurs (for instance, as the effect of an impact or simple aging). The area of interests should be then regarded as a "hot spot", worthy of monitoring along the system lifetime. In this case, the inclusion of embedded fibers shall be relevant for a continuous strain levels monitoring under operational loads at ground (a very basic, but perhaps effective SHM system).

As experimental campaigns should be designed very carefully, in order to properly characterize the phenomenon under investigation, in the same way numerical investigations activity could be structured and detailed in order to attain specific information and possibly, relevant solutions.

Author Contributions: Conceptualization, S.A., I.D. and L.P.; methodology, S.A., I.D. and A.C.; software, S.A.; validation, A.C. and U.M.; formal analysis, L.P.; investigation, L.P. and I.D.; resources, U.M.; data curation, L.P.; writing—original draft preparation, S.A.; writing—review and editing, I.D. and A.C.; visualization, A.C.; supervision, U.M.; project administration, U.M.; funding acquisition, U.M. All authors have read and agreed to the published version of the manuscript.

Funding: This research was funded by Clean Sky 2 Joint Undertaking, under the European's Union Horizon 2020 research and innovation Programme, under grant agreement No 807089—REG GAM 2018—H2020-IBA-CS2- GAMS-2017.

Institutional Review Board Statement: Not applicable.

Informed Consent Statement: Not applicable.

Data Availability Statement: Not applicable.

Acknowledgments: The AirGreen2 Project has received funding from the Clean Sky 2 Joint Undertaking, under the European's Union Horizon 2020 research and innovation Programme, under grant agreement No 807089—REG GAM 2018—H2020-IBA-CS2- GAMS-2017.

Conflicts of Interest: The authors declare no conflict of interest.

Abbreviations

A/C	Aircraft
CFD	Computational Fluid Dynamics
CS2	Clean Sky 2 Programe
CFRP	Carbon Fiber Reinforced Polymer
CL	Lift coefficient
DOF	Degree of Freedom
FE	Finite Element
FEM	Finite Element Model
GRA	Green Regional Aircraft
LCA	Load Control and Alleviation
LDO	Leonardo Aircraft Division
LE	Leading Edge
LoD	Lift over Drag
RL	Standard Reference Load
SDOF	Single Degree of Freedom
SHM	Structural Health Monitoring
TE	Trailing Edge
UD	Unidirectional

References

1. Ameduri, S.; Concilio, A.; Dimino, I.; Pecora, R.; Ricci, S. AIRGREEN2—Clean Sky 2 Programme: Adaptive Wing Technology Maturation, Challenges and Perspectives. In Proceedings of the ASME 2018 Conference on Smart Materials, Adaptive Structures and Intelligent Systems, San Antonio, TX, USA, 10–12 September 2018. [CrossRef]
2. Peter, F.; Stumpf, E. The development of morphing aircraft benefit assessment. In *Morphing Wing Technologies, Large Commercial Aircraft and Civil Helicopters*; Concilio, A., Dimino, I., Lecce, L., Pecora, R., Eds.; Butterworth-Heinemann: Oxford, UK, 2018; Volume 3, pp. 103–121.
3. Kota, S.; Osborn, R.; Ervin, G.; Maric, D. Mission adaptive compliant wing–design, fabrication and flight test. In Proceedings of the NATO Research Technology Organization Research Symposium, Morphing Vehicles, RTO-MP-AVT-168 2009, Evora, Portugal, 20–24 April 2009.
4. NASA; US AFRL. Adaptive compliant trailing edge flight experiment. *RC Soar. Digest* **2014**, *31*, 85–86.
5. Herrera, C.Y.; Spivey, N.D.; Lung, S.F.; Ervin, G.; Arbor, A.; Flick, P. Aeroelastic airworthiness assessment of the adaptive compliant trailing edge flaps. In Proceedings of the 46th Society of Flight Test Engineers International Symposium, Lancaster, CA, USA, 14–17 September 2015.
6. Radestock, M.; Riemenschneider, J.; Monner, H.P.; Rose, M. Experimental investigation of a compliant mechanism for an uav leading edge. In Proceedings of the 7th Eccomas Thematic Conference on Smart Structures and Materials, Smart 2015, Azores, Portugal, 3–6 June 2015.
7. Arena, M.; Amoroso, F.; Pecora, R.; Amendola, G.; Dimino, I.; Concilio, A. Numerical and experimental validation of a full scale servo-actuated morphing aileron model. *Smart Mater. Struct.* **2018**, *27*, 105034. [CrossRef]
8. McGowan, A.-M.R.; Vicroy, D.D.; Busan, R.C.; Hahn, A.S. Perspectives on Highly Adaptive or Morphing Structures. In Proceedings of the NATO Research Technology Organization Applied Vehicle Technology Panel Symposium, RTO-MP-AVT-152 2009, Evora, Portugal, 20–24 April 2009.
9. Chattopadyay, N.C.; Jony, B.; Acharya, A. An analysis on wing morphing. In Proceedings of the Global Engineering, Science and Technology Conference 2012, Dhaka, Bangladesh, 28–29 December 2012.
10. Delery, J.; Fulker, J.; de Matteis, P. *Drag Reduction by Shock and Boundary Layer Control: Results of the Project Euroshock II, Supported by the European Union 1996–1999, Notes on Numerical Fluid Mechanics and Multidisciplinary Design*; Springer Science & Business Media: Berlin/Heidelberg, Germany, 2013; Volume 80, ISBN 978-3-540-45856-2.
11. Stanewsky, E.; Delery, J.; Fulker, J.; Geissler, W. Euroshock-Drag Reduction by Passive Shock Control: Results of the Project Euroshock, AER2-CT92–0049, Supported by the European Union, 1993–1995. In *Notes on Numerical Fluid Mechanics*; Springer Science & Business Media: Berlin/Heidelberg, Germany, 2013; Volume 56, ISBN 978-3-322-90711-0.
12. Ameduri, S.; Concilio, A. Morphing wings review: Aims, challenges, and current open issues of a technology. *Proc. Inst. Mech. Eng. Part C J. Mech. Eng. Sci.* **2020**. [CrossRef]

13. Fleming, G.A.; Burner, A.W. Deformation measurements of smart aerodynamic surfaces. In Proceedings of the SPIE Conference on Optical Diagnostics for Fluids/Heat/Combustion and Photomechanics for Solids, Denver, CO, USA, 21–23 July 1999.

14. Ameduri, S.; Ciminello, M.; Concilio, A.; Brindisi, A.; Petrella, O.; Tiseo, B.; Mazzola, L.; Piscitelli, F.; Sorrentino, R. Design approach of a large strain sensor based on nanoparticle technology: A highly-integrable sensor for Morphing applications including SHM & shape reconstruction. In Proceedings of the 2017 13th Conference on Ph.D. Research in Microelectronics and Electronics (PRIME), Taormina, Italy, 12–15 June 2017.

15. Romano, F.; Barile, M.; Cacciapuoti, G.; Godard, J.L.; Vollaro, P.; Barabinot, P. Advanced OoA and Automated Technologies for the Manufacturing of a Composite Outer Wing Box. In Proceedings of the 8th EASN-CEAS International Workshop on Manufacturing for Growth and Innovation, Glasgow, UK, 4–7 September 2018; pp. 1–8. [CrossRef]

16. Available online:. Available online: https://www.compositesworld.com/articles/multilayer-thermoplastic-tapes-afp-and-resin-infusion-for-more-democratic-composites (accessed on 13 March 2021).

17. Moens, F. Augmented Aircraft Performance with the Use of Morphing Technology for a Turboprop Regional Aircraft Wing. *Biomimetics* **2019**, *4*, 64. [CrossRef] [PubMed]

18. de Gaspari, A.; Cavalieri, V.; Ricci, S. Advanced Design of a Full-Scale Active Morphing Droop Nose. *Int. J. Aerosp. Eng.* **2020**, *2020*, 1086518. [CrossRef]

19. Rea, F.; Amoroso, F.; Pecora, R.; Noviello, M.C.; Arena, M. Structural design of a multifunctional morphing fowler flap for a twin-prop regional aircraft. In Proceedings of the ASME 2018 Conference on Smart Materials, Adaptive Structures and Intelligent Systems, San Antonio, TX, USA, 10–12 September 2018. [CrossRef]

20. Fonte, F.; Toffol, F.; Ricci, S. Design of a Wing Tip Device for Active Maneuver and Gust Load Alleviation. In Proceedings of the AIAA/ASCE/AHS/ASC Structures, Structural Dynamics, and Materials Conference—AIAA SciTech, Kissimmee, FL, USA, 8–12 January 2018; pp. 1–18, ISBN 9781624105326. [CrossRef]

21. Amendola, G.; Dimino, I.; Concilio, A.; Pecora, R.; Cascio, M.L. Preliminary Design Process for an Adaptive Winglet. *Int. J. Mech. Eng. Robot. Res.* **2018**, *7*. [CrossRef]

22. Dimino, I.; Amendola, G.; Di Giampaolo, B.; Iannaccone, G.; Lerro, A. Preliminary design of an actuation system for a morphing winglet. In Proceedings of the 8th International Conference on Mechanical and Aerospace Engineering (ICMAE), Prague, Czech Republic, 22–25 July 2017; pp. 416–422. [CrossRef]

23. Dimino, I.; Andreutti, G.; Moens, F.; Fonte, F.; Pecora, R.; Concilio, A. Integrated Design of a Morphing Winglet for Active Load Control and Alleviation of Turboprop Regional Aircraft. *Appl. Sci.* **2021**, *11*, 2439. [CrossRef]

24. Dimino, I.; Gallorini, F.; Palmieri, M.; Pispola, G. Electromechanical Actuation for Morphing Winglets. *Actuators* **2019**, *8*, 42. [CrossRef]

25. Noviello, M.C.; Dimino, I.; Concilio, A.; Amoroso, F.; Pecora, R. Aeroelastic Assessments and Functional Hazard Analysis of a Regional Aircraft Equipped with Morphing Winglets. *Aerospace* **2019**, *6*, 104. [CrossRef]

Article

A Rational Numerical Method for Simulation of Drop-Impact Dynamics of Oleo-Pneumatic Landing Gear

Rosario Pecora

Industrial Engineering Department, Aerospace Division, University of Naples "Federico II", 80125 Napoli, Italy; rosario.pecora@unina.it

Abstract: Oleo-pneumatic landing gear is a complex mechanical system conceived to efficiently absorb and dissipate an aircraft's kinetic energy at touchdown, thus reducing the impact load and acceleration transmitted to the airframe. Due to its significant influence on ground loads, this system is generally designed in parallel with the main structural components of the aircraft, such as the fuselage and wings. Robust numerical models for simulating landing gear impact dynamics are essential from the preliminary design stage in order to properly assess aircraft configuration and structural arrangements. Finite element (FE) analysis is a viable solution for supporting the design. However, regarding the oleo-pneumatic struts, FE-based simulation may become unpractical, since detailed models are required to obtain reliable results. Moreover, FE models could not be very versatile for accommodating the many design updates that usually occur at the beginning of the landing gear project or during the layout optimization process. In this work, a numerical method for simulating oleo-pneumatic landing gear drop dynamics is presented. To effectively support both the preliminary and advanced design of landing gear units, the proposed simulation approach rationally balances the level of sophistication of the adopted model with the need for accurate results. Although based on a formulation assuming only four state variables for the description of landing gear dynamics, the approach successfully accounts for all the relevant forces that arise during the drop and their influence on landing gear motion. A set of intercommunicating routines was implemented in MATLAB® environment to integrate the dynamic impact equations, starting from user-defined initial conditions and general parameters related to the geometric and structural configuration of the landing gear. The tool was then used to simulate a drop test of a reference landing gear, and the obtained results were successfully validated against available experimental data.

Keywords: landing gear dynamics; drop impact; oleo-pneumatic shock absorber; ground loads; spin-up; spring-back

Citation: Pecora, R. A Rational Numerical Method for Simulation of Drop-Impact Dynamics of Oleo-Pneumatic Landing Gear. *Appl. Sci.* **2021**, *11*, 4136. https://doi.org/10.3390/app11094136

Academic Editor: Ruxandra Mihaela Botez

Received: 26 February 2021
Accepted: 29 April 2021
Published: 30 April 2021

1. Introduction

Oleo-pneumatic landing gear represents a key technology in modern aviation due to its remarkable efficiency in smoothly absorbing large amounts of energy in a short period of time. The rapid growth of airplane dimensions, weight, and flight speed has been made possible by the development of ever more sophisticated systems enabling safe landing throughout the entire operative life of the vehicle. Owing to its compact dimensions, relatively competitive weight, and fatigue life, the oleo-pneumatic strut has quickly replaced the classical cantilever solution (Figure 1a), which is nowadays only considered for the main landing units of very light aircraft or small general aviation aircraft [1].

Apart from the general layout, the main difference between cantilever and oleo-pneumatic struts is in the way the landing energy is dissipated. In the first case, the dissipation occurs through the deformation of an elastic beam connecting the wheels' axles to the fuselage; in the second case, it occurs through the motion of a piston inside a cylinder filled with a mixture of gas and oil (Figure 1b).

Appl. Sci. **2021**, *11*, 4136. https://doi.org/10.3390/app11094136　　　　https://www.mdpi.com/journal/applsci

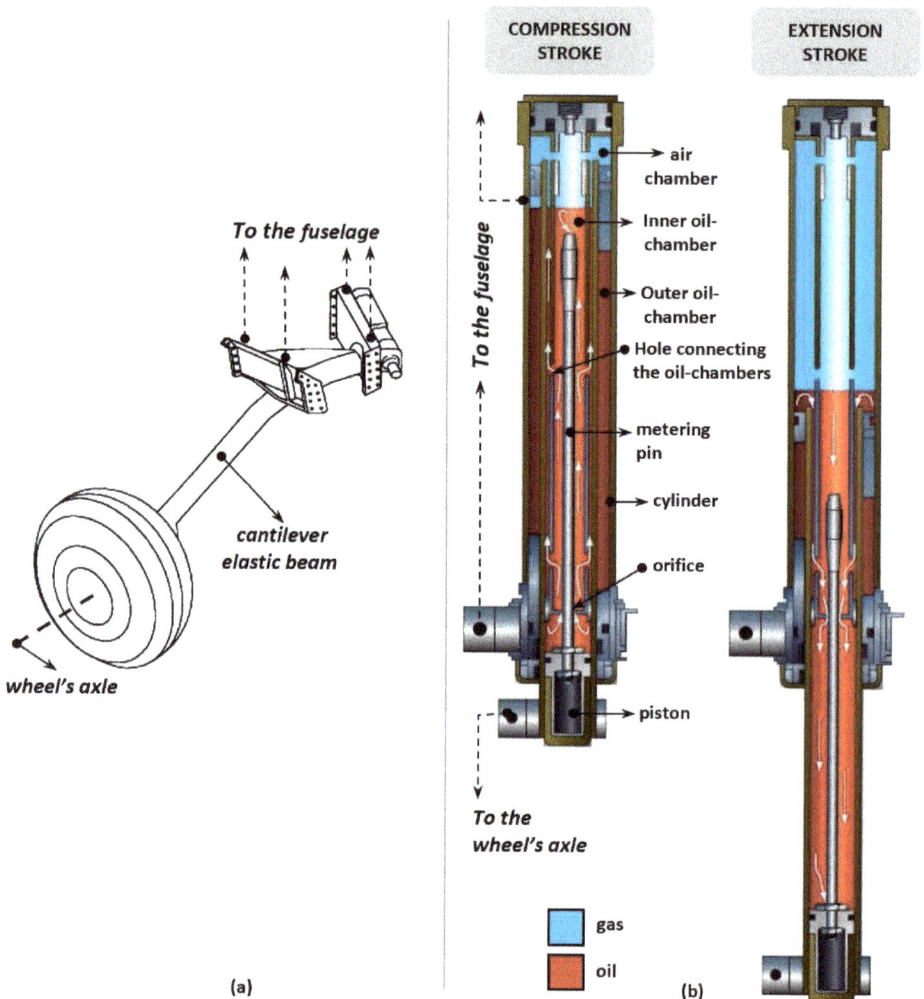

Figure 1. Main types of landing gear: (**a**) cantilever strut [2] and (**b**) oleo-pneumatic shock-absorber [3].

The piston-cylinder assembly is generally referred to as a *shock absorber*, and the dissipative force generated during the landing is mainly due to the flow of viscous oil into an orifice with a cross-section that varies with the piston stroke. In more detail, the oil flows between two coaxial chambers, depending on the direction of motion of a tapered metering pin mounted on the piston (Figure 1b). During the compression stroke, the metering pin enters the inner chamber through an orifice with net section area A_s; to compensate for the reduction in the amount of available volume, part of the oil contained in the inner chamber moves into the outer chamber, flowing into the same orifice crossed by the metering pin and into some holes connecting the two chambers. At the same time, the gas contained in a compensation air chamber (placed at the top of the oil chamber) is compressed, thus contributing to the balance of fluid volume in the different compartments of the cylinder.

Unlike the communication holes between oil chambers, the portion of the orifice hosting the oil flow has a variable section area, A_{tr}, given by the difference $A_S–A_{mp}$, where A_{mp} is the metering pin's net section area, which varies with the pin length. During the extension stroke, the pin comes out the inner oil chamber, the air chamber volume expands, and the oil flows from the outer chamber into the inner chamber. The damping force produced by the oleo-pneumatic system can be then modulated by working on the shock absorber's main characteristics, with the ultimate goal of adequately limiting the aircraft's vertical acceleration to the maximum value the aircraft structure can absorb after impacting the ground.

A reliable simulation tool for landing gear impact dynamics can therefore provide tremendous support for the design of new airplanes, as it facilitates the integrated definition of coherent layouts for both the landing unit and the airframe. In full awareness of this consideration, several approaches have been proposed to efficiently model the dynamic behavior of oleo-pneumatic landing gear, each with its own level of complexity, pros and cons.

The first attempt at shock absorber analysis was made by Hadekel [4], who focused on evaluating the system's overall damping efficiency rather than proposing effective models for simulating the impact dynamics. Milwitzky and Cook [5] presented an integral model for the entire shock absorber, which was considered as a dynamic system with a single degree of freedom. The model was successfully validated by experiments but did not accurately account for the influence of many parameters involved in the drop dynamics. Currey [6] investigated the relationship between the dissipated energy and the air chamber's physical characteristics and described effective thermodynamic models to predict the behavior of the gas during the compression and extension stroke of the piston. Relying on these studies, Daniels [7] built a sophisticated model allowing for precise simulation of hydraulic dissipative forces; unfortunately, its validity was limited to telescopic landing gears only.

Working on a six degree-of-freedom system, Sivakumar and Haran [8,9] were able to fairly represent shock absorber dynamics during taxiing. Their approach adopted linear differential equations of motion, which were very useful in predicting the vibration levels induced by the ground asperities on the fuselage structure and passengers. On the other hand, the equations could not realistically capture the energy dissipation mechanism at touchdown due to their linear formulation. To overcome this limitation, Sivakumar and Syedhaleem [10] used a series of two mass-spring-damper models to simulate the impact dynamics of the shock absorber and wheel assembly. The model was validated against the numerical results of finite element analysis carried out in ABAQUS [11] and proved to be reliable in predicting the time history of vertical displacement, acceleration, and reaction force. Unfortunately, no equations were considered for obtaining the wheel axle's vibrations along the horizontal direction consequent to the aircraft flight speed at the moment of impact, and the friction between the tire and the ground. This made it impossible to estimate spin-up and spring-back loads, which are among the most severe loads in landing gear design [12].

Recent studies offer comprehensive simulation strategies for landing gear impact dynamics, relying upon ever more advanced finite element solvers; some relevant examples can be found in a good review article by Kruger and Morandini [13].

An interesting finite element analysis accounting for the nonlinear problems in landing dynamics was proposed by Lyle [14], while Khapane et al. used multibody dynamics software (Simpack) combined with the FE solver NASTRAN to investigate the load instability produced by the interaction of the brake mechanism with the landing gear's wheels [15]. Lernbeiss introduced a multi-body system-based landing gear model for numerical simulation of simple static and dynamic load conditions and successfully correlated the simulation results with the outcomes of a more accurate finite element simulation [16]. The Lernbeiss model was then revised and improved by Kong, based on drop-impact analysis and experiments conducted on a smart UAV landing system [17]. LMS® analysis software for multi-body dynamics was used by Xue and Yu to build a coupled rigid-flexible simulation model of light aircraft main landing gear, showing very good capability in terms of drop-test load estimation [18].

Finite element models represent effective tools to support the design of landing gear, enabling accurate investigation of the mechanical behavior of each subcomponent under the action of static and dynamic loads. The development of new software for parametric FEA and the introduction of elastic models in simplified multi-body dynamics packages have recently led to the possibility of accounting for some enhanced levels of detail in simulations of drop dynamics even in the preliminary design stage of the landing gear, while also allowing for trade-off studies on the effects induced by configuration changes on the overall landing performance. On the other hand, FE-based approaches become somewhat impractical at the beginning of the preliminary design phase of an aircraft, when there are too many undefined design variables to consider the generation of the landing gear's 3D CAD or structural mesh convenient for estimating its influence on ground loads.

Referring to modern aircraft, characterized by ever more light and flexible structures, the possibility of predicting ground loads and vertical acceleration levels at landing is crucial from the early stages of the project, as it enables proper design choices to be made.

According to airworthiness regulations, determining ground loads and acceleration follows a general process that is independent of the aircraft certification category and can be summarized in four consecutive steps:

1. Estimate the design load factor at landing by means of conservative assumptions or rational analysis. (The design load factor at landing, or *ground load factor*, is defined as the ratio between the maximum vertical acceleration arising at the aircraft's center of gravity during landing and the acceleration of gravity.)
2. Apply standard criteria to evaluate landing loads and vertical acceleration referring to the assumed load factor and design-relevant landing conditions defined in terms of the aircraft's inertial configurations and attitudes.
3. Determine the most critical load/acceleration and landing conditions among those of the previous step.
4. After landing system design and manufacture, conduct drop tests on landing gear and experimentally evaluate the maximum ground load factor.

The reference aircraft certification category has an influence only on the specific approaches and formulations that must be followed to accomplish the tasks at each step, while the general logic of the process is essentially common to all certification standards.

If the experimental load factor obtained by drop tests is lower than that assumed at the beginning of the process, then the landing gear design is approved, provided that it is also shown (by test) that the landing gear is also able to withstand the most critical static/dynamic load conditions among those evaluated.

Landing gear design approval is-therefore-based on demonstrating compliance with two concomitant criteria:

a. The structure must be able to withstand the most severe loads expected in service without damage or failure (as for all components of the aircraft primary structure).
b. The maximum ground load factor developed during prescribed drop tests must be lower than that assumed for the evaluation of ground loads in the first place.

If the landing gear fails to meet the first criterion, then recovery actions can be taken to strengthen the structural layout of the item and new tests can be performed to prove the adequacy of the implemented corrections. If it fails to meet the second criterion, then the entire set of ground loads is compromised and needs to be evaluated again, in correspondence of a new assumption on the load factor value; this severely impacts the design process of the landing gear and the structural components whose layout is influenced by ground loads magnitude.

The initial estimation of ground load factor therefore plays an important role in the correct assessment of ground loads and the subsequent design of the airframe and landing system; the accuracy of this estimation is ultimately related to good prediction of the landing gear's shock absorption performance in standardized drop tests, and at the phase of the project when the design of the landing gear has just started and no detailed simulation models are available.

In line with this evidence, the author developed a numerical method for simulating landing gear dynamics in a typical certification drop test, which rationally balances the low level of sophistication of the adopted model with the need for accurate results. A nonlinear dynamic model is adopted, which properly accounts for all the energy-dissipation mechanisms occurring at the impact; in this regard, the shock absorber's vertical reaction force is combined with the horizontal solicitation induced by the tire friction with the ground. A general description of the proposed model is outlined in this work, together with the adopted formulation and software tools to simulate the impact dynamics.

Relying upon these tools, the drop test of a reference landing gear was numerically replicated, and the implemented approaches were successfully validated against already available experimental data.

2. Formulation of Landing Gear Impact Dynamics

The landing gear was modeled as an elastic system with four degrees of freedom, adequately selected to effectively capture the impact dynamics.

The following Lagrangian coordinates were defined with reference to the Cartesian system S depicted in Figure 2:

- x_1, z_1: x- and z-coordinates of point M located at the center of the wheel hub;
- z_2: z-coordinate of the cylinder head; and
- φ: wheel rotation angle.

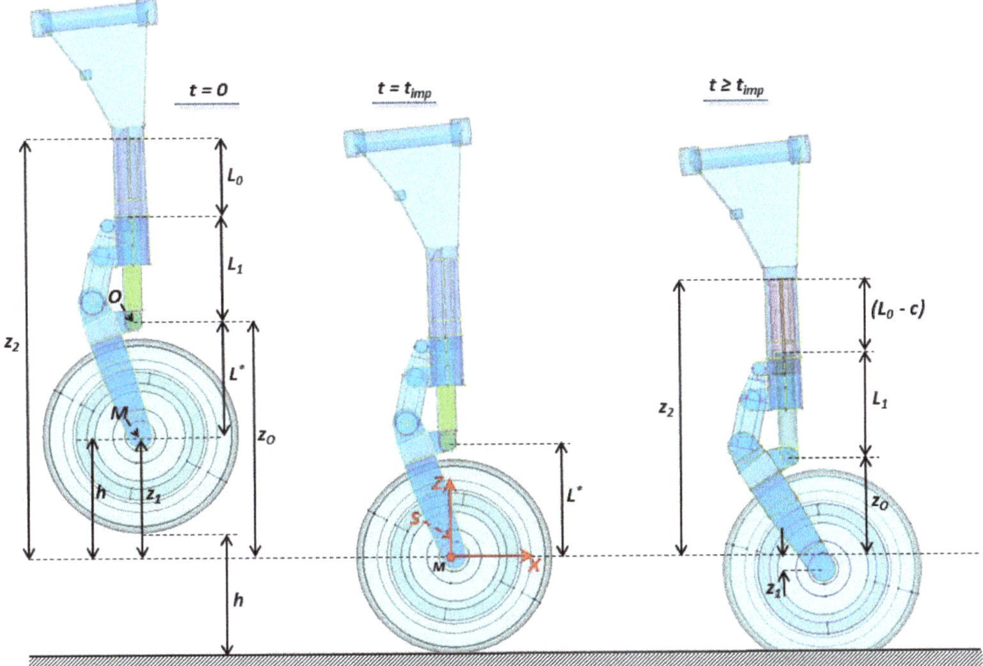

Figure 2. Phases of landing gear drop and main parameters involved in impact dynamics.

According to the scheme of Figure 2, the impact dynamics was roughly divided into three subsequent time phases:

1. Free fall, on the left side of the figure, during which the landing gear is approaching the ground;
2. Impact, at the center of the figure, occurring at the single instant t_{imp}, when the wheel touches the ground; and
3. Shock absorption, on the right side of the figure, during which the landing gear works to dampen the impact-induced acceleration.

If c indicates the stroke of the piston and z_0 the z-coordinate of point O located at the base of the piston corresponding to the piston–fork joint, at any instant of time t, the result is:

$$z_O(t) + L_1 + L_0 - c(t) = z_2(t) \tag{1}$$

where L_0 and L_1 indicate the longitudinal dimension of the air chamber and piston, respectively.

Equation (1) can be easily rearranged to express the stroke as a function of variables z_0 and z_2:

$$c(t) = z_O(t) - z_2(t) + L_1 + L_0 \tag{2}$$

Although not evident in Equation (2), the stroke is dependent only on Lagrangian coordinates z_1 and z_2 ($c(t) = c(z_1(t), z_2(t))$). z_0 is a dummy variable used only to express the stroke in a more intuitive and compact manner; this variable is linked to z_1 and z_2 by a nonlinear equation, which is derived in Appendix A by means of simple geometric and kinematic considerations.

During free fall, the stroke is identically equal to zero.

At the origin of the investigation time frame ($t = 0$), we have:

$z_1(0) = h$, where h is the drop height,
$z_2(0) = h + L^* + L_1 + L_0$, where L^* is the vertical distance between points O and M when $t \leq t_{imp}$,
$z_0(0) = h + L^*$,

and, according to Equation (2):

$$c(0) = z_O(0) - z_2(0) + L_1 + L_0 = h + L^* - h - L^* - L_1 - L_0 + L_1 + L_0 = 0.$$

If $0 < t < t_{imp} = \sqrt{2h/g}$, then:

$$z_1(t) = z_1(0) - \tfrac{1}{2}gt^2,$$
$$z_2(t) = z_2(0) - \tfrac{1}{2}gt^2,$$
$$z_0(t) = z_0(0) - \tfrac{1}{2}gt^2,$$

and, again,

$$c(t) = z_O(0) - \tfrac{1}{2}gt^2 - z_2(0) + \tfrac{1}{2}gt^2 + L_1 + L_0 = h + L^* - h - L^* - L_1 - L_0 + L_1 + L_0 = 0.$$

The stroke is also identically null at the instant of impact; when $t = t_{imp} = \sqrt{2h/g}$, the following occurs:

$$z_1(t_{imp}) = 0,$$
$$z_2(t_{imp}) = L^* + L_1 + L_0,$$
$$z_0(t_{imp}) = L^*,$$

and, finally,

$$c(t_{imp}) = z_O(t_{imp}) - z_2(t_{imp}) + L_1 + L_0 = L^* - L^* - L_1 - L_0 + L_1 + L_0 = 0.$$

When $t > t_{imp}$, the piston's stroke is no longer equal to zero; the motion of the piston into the cylinder generates vertical reaction forces F_a and F_d due to the compression of gas and the flow of oil into the oleo-pneumatic chamber, respectively. In addition, a further reaction force, F_{PN}, arises along the vertical direction due to the compression of gas in the tire. F_a and F_d depend on the geometric characteristics of the oleo-pneumatic chamber and can be expressed as functions of the piston's stroke and its time derivative, according to the following equations [7]:

$$F_a = F_a(c) = P_0 A_0 [L_o/(L_o - c)]^\gamma \tag{3}$$

$$F_d = F_d(\dot{c}) = \rho_{oil} A_s^3 \dot{c}|\dot{c}|/(2A_{tr}^2 c_{visc}^2) \tag{4}$$

where A_0 and P_0 are the air chamber section area and pre-load pressure in the air chamber, respectively; L_0 is the initial length of the air chamber (i.e., its longitudinal length at $t < t_{inp}$); γ is the polytropic index of gas in the air chamber; ρ_{oil} and c_{visc} are the density and viscosity of oil in the hydraulic chamber, respectively; A_s is the section area of the piston; and A_{tr} is the net section of the orifice of the hydraulic chamber traversed by the oil flow, generally variable with the piston stroke.

F_{PN} basically depends on the tire's mechanical properties and is generally provided by tire manufacturers in the form of experimental curves relating the reaction force to the tire crushing and inflation pressure. For a given tire and inflation pressure, F_{PN} is therefore considered as a function of the variable z_1 only. The tire's overall reaction force is more precisely given by the sum of F_{PN} and F_{d0}, where F_{d0} stands for the reaction force due to the deformation of the tire's material (at zero inflating pressure). F_{d0} is dependent on the rate of deformation of the tire (dz_1/dt) since the tire's material is characterized by viscoelastic behavior. Within the limits of the proposed formulation, F_{d0} was considered negligible compared to F_{PN}.

By imposing the equilibrium of all forces acting along the z-axis of the reference S (Figure 2), the following equations for the impact dynamics of the system can be written:

$$\begin{cases} m\ddot{z}_1 &= -F_a(c) - F_d(\dot{c}) + F_{PN}(z_1) - mg \\ M_e\ddot{z}_2 &= F_a(c) + F_d(\dot{c}) - M_e g \end{cases} \tag{5}$$

where M_e is the effective mass, supposed to be lumped at the head of the cylinder and accounting for the amount of aircraft weight acting on the landing gear element at the moment of impact, and m is the swinging mass, supposed to be lumped at the middle point of the wheel hub and equal to the sum of the wheel and tire masses, plus one-third of the landing gear leg's mass.

Equation (5) are valid also when $t < t_{imp}$, and basically describe the free fall of masses M_e and m as two separate bodies;in this case F_a, F_d and F_{PN} are all equal to zero.

After the impact, the elastic oscillation of the wheel hub along the x-axis is governed by the equation:

$$m\ddot{x}_1 = -\sigma \dot{x}_1 - Kx_1 - \frac{\dot{s}}{|\dot{s}|} F_x \tag{6}$$

where σ and K are the structural (viscous) damping and stiffness exhibited by the landing gear leg along the x-axis; F_x is the module of the friction force due to the tire contact with the ground, which can be reasonably assumed equal to $\mu \cdot F_{PN}(z_1)$, where μ is the tire sliding coefficient of friction, and s is the x-displacement of the point of contact between the tire and the ground (tire slip, Figure 3), given by $(x_1 + \varphi\rho)$, where ρ is the distance of the wheel hub axis to the ground, or the difference between the radius of the wheel and the tire crushing.

The friction force F_x and displacement s do not represent two independent variables, provided that they are both called to satisfy the rotational equilibrium equation of the wheel about the hub axis. Under the assumption of negligible bearing friction, the wheel rotational equilibrium may be formulated in the following way:

$$I\ddot{\varphi} = \rho F_x \tag{7}$$

with I equal to the wheel's polar moment of inertia. Well-performing, and well-lubricated, bearings are commonly used in aeronautics; consequently, and within the limits of approximation of the proposed formulation, the torque due to bearing friction can be reasonably considered negligible compared to that induced by the friction force between the tire and the ground.

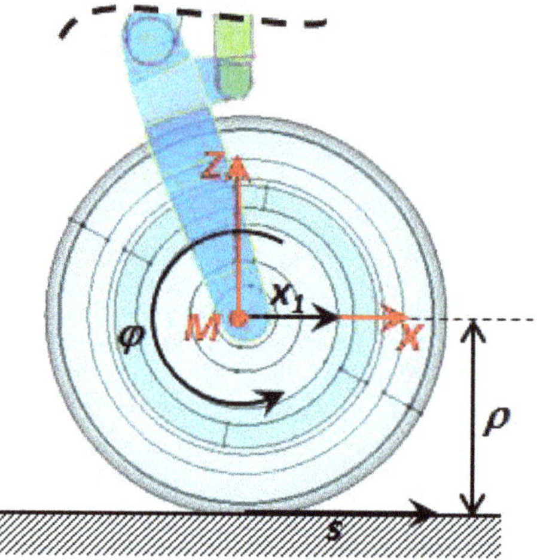

Figure 3. Graphical sketch explaining relationships among variables s, x_1, and φ.

As a whole, Equations (5)–(7) define the system of nonlinear and coupled differential equations whose solution describes the impact dynamics of the landing gear in terms of the time-dependent state variables x_1, z_1, z_2, and φ.

It is worth noting that the expressions of the stroke given by (2) and the vertical equilibrium given by (5) are valid for the specific landing gear arrangement depicted in Figure 2.

This arrangement is typical of main landing gear and is characterized by a hinged connection between the fork, the piston, and the cylinder, usually referred to as a *trailing link*. The cylinder's inclination angle with respect to the vertical axis z is null or negligible. In some other arrangements generally adopted for nose landing gear, the fork and piston move together with no relative rotation between them, and the movement occurs along a direction that is inclined with respect to the normal to the ground (Figure 4). In such cases, Equations (2) and (5) can be easily rewritten following the same approach that has been used so far.

Referring to the typical nose landing gear arrangement in Figure 4 (telescopic landing gear), we can immediately observe that:

$$z_1(t) + L_1 + L_0 - c(t) \cdot cos\theta = z_2(t)$$

and then,

$$c(t) = c((z_1(t), z_2(t)) = (z_1(t) - z_2(t) + L_1 + L_0)/cos\theta. \tag{8}$$

By imposing the equilibrium along the vertical axis z, we can finally obtain:

$$\begin{cases} m\ddot{z}_1 &= [-F_a(c) - F_d(\dot{c})]cos\theta + F_{PN}(z_1) - mg \\ M_e\ddot{z}_2 &= [F_a(c) + F_d(\dot{c})]cos\theta - M_eg \end{cases}. \tag{9}$$

Equations (6) and (7) have general validity and can therefore be used together with (8) and (9) to complete the set of equations governing the impact dynamics of the landing gear arrangement of Figure 4.

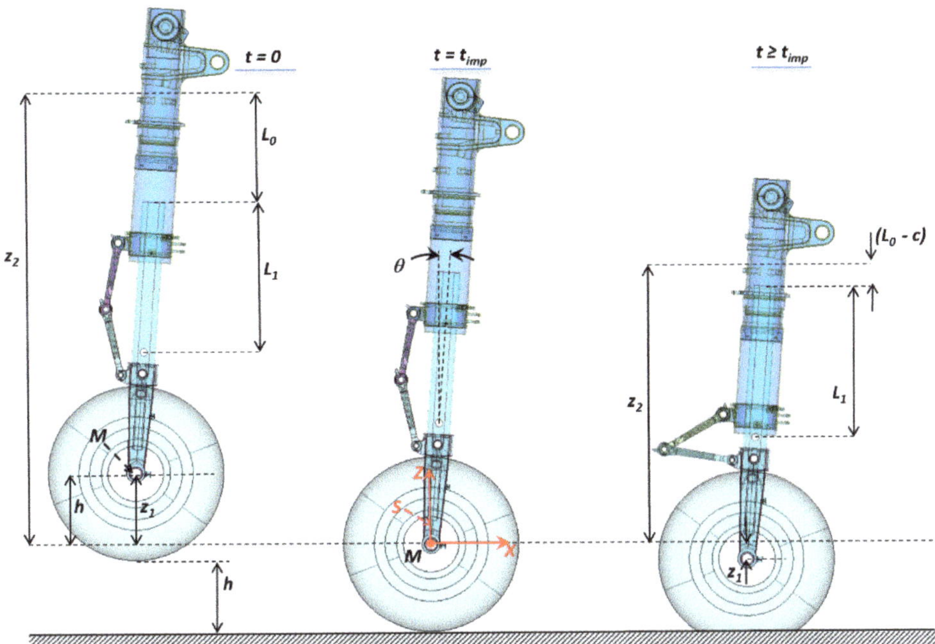

Figure 4. Phases of landing gear drop and main parameters involved in impact dynamics (telescopic arrangement, typical of nose landing gear).

In conclusion, the proposed simulation approach leads to two different formulations for the equations of drop impact dynamics, depending on the type of landing gear under investigation. Equations (5)–(7), and the expression of the stroke given by (2) apply to landing gear with a trailing link arrangement and vertical leg (typical configuration of main landing systems), while Equations (6), (7), and (9) combined with Equation (8) for the stroke apply to nose landing gear, generally characterized by the configuration depicted in Figure 4.

During the preliminary landing gear design stage, a geometric layout of the item is assumed in combination with a wheel type. Based on these assumptions and the drop-test prescriptions of the reference airworthiness regulations, all the input data required to run the simulation are available at the preliminary design stage, with the exception of σ and K (used in Equation (6)).

Nevertheless, first trial values may be assigned to σ and K by using the structural similitude of landing gear belonging to the same category (i.e., suitable for installation on aircraft in the same category of MTOW). It can be observed that the natural frequency of the (main or nose) landing gear's fore and aft bending mode does not significantly change over different landing gear models in the same category.

If the natural frequency f of the fore-and-aft bending mode is reasonably assumed according to the landing gear category, then K can be approximately obtained as $m*4\pi*f^2$, with m equal to the mass of the wheel; similarly, σ can be expressed as $\zeta*\sigma_{cr}$, where σ_{cr} is the critical damping given by $2\,m*(K/m)^{0.5} = 4\,m\pi f$ and ζ is the damping ratio, which is generally in the range of 0.03–0.05 for metallic structures with joints.

The design of the landing gear is clearly an iterative process; after the preliminary assumptions of σ, K (and on all other input data needed for the simulation), the first loop of dynamic impact loads can be evaluated and a preliminary structural model can be generated for the landing gear. New simulations can then be carried out to produce the second loop of loads, this time using input values directly derived from the preliminary structural model. The second loop of loads is used to update the structural model, and the process goes on until the executive design of the item is reached. In all of these stages, the proposed simulation approach provides an efficient tool to evaluate the drop dynamics of the landing gear (and associated loads), relying upon input data that are continuously updated to follow the progress of the design.

3. Drop Test Simulation and Results

A set of intercommunicating MATLAB® [19] routines was implemented and arranged in a compact software environment to automatically solve the differential equations governing the drop dynamics of the landing gear according to the approaches described in Section 2.

The tool consists of three main modules (Figure 5):

- An input module, which reads and processes the input data of the problem, in terms of both the landing gear type (main or nose) and characteristics.
- A solver module, which implements the Runge-Kutta method [20] to integrate the system of differential equations numerically.
- An output module, which prints the time histories of the most relevant physical parameters and plots practical diagrams to visually support the result analysis.

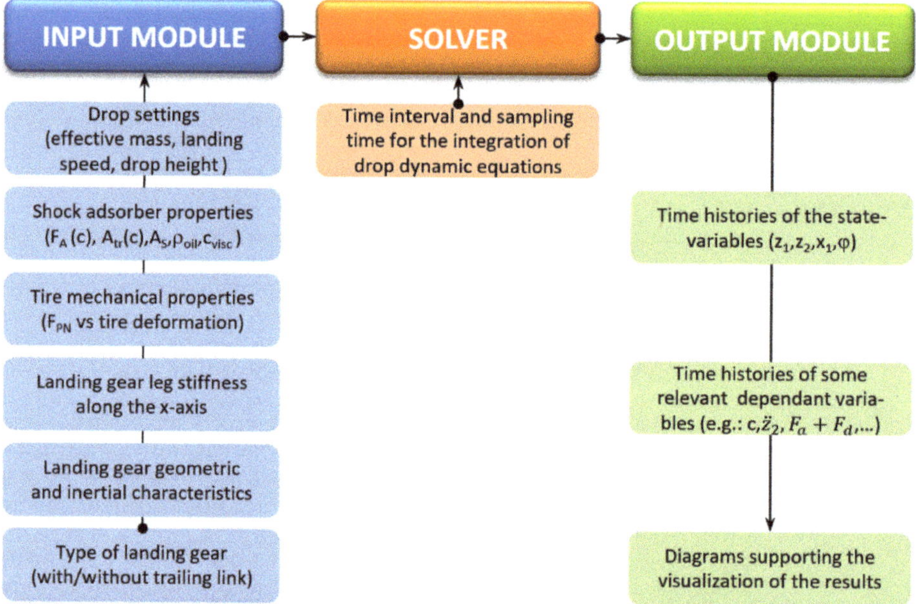

Figure 5. Main modules and structure of drop test simulation tool.

To check the reliability of the proposed numerical approach and its implementation in the software tool, a drop test of a reference landing gear was simulated, and the results were compared with already available experimental data.

The main landing gear of a large UAV was selected as a reference; the name and pictures of the aircraft are here omitted for confidentiality reasons. To properly frame the application scenario, it is sufficient only to mention that the UAV had a maximum landing weight (*MLW*) of 3040 kg and a wing surface area (*S*) of 21 m^2, and was designed in compliance with NATO-STANAG 4671 airworthiness requirements [21].

Among the tests performed to qualify the main landing gear, the limit drop (paragraph 725 of [21]) was selected as a benchmark case for the drop simulation code.

According to the limit drop test requirements, the drop height must be equal to:

$$h = 0.0132 \sqrt{\frac{MLW_{[\text{Kg}]} \cdot g_{[\frac{m}{s^2}]}}{S_{[\text{m}^2]}}} \tag{10}$$

and not lower than 0.234 m or greater than 0.475 m.

The landing gear must be dropped while carrying an effective mass equal to:

$$M_e = M\frac{h + (1 - L)d}{h + d} \tag{11}$$

where M_e is the effective mass to be used in the drop test; h is the specified drop height; d is the deflection under the impact of the tire, plus the vertical component of the wheel hub travel relative to the drop mass at the instant of time when the piston stroke reaches its maximum value; M is equal to the static weight on the main landing gear with the UAV in level attitude with the nose wheel clear; and L is the ratio of assumed wing lift to UAV weight (not greater than 2/3).

Replacing $MLW = 3040$ kg and $S = 21$ m^2 in Equation (10), the reference landing gear's drop height was equal to 0.497 m and therefore was limited to 0.475 m per regulatory requirements.

The effective mass to be used in the drop test depends on the parameter d, which is an output of the drop itself; for this reason, an iterative process, based on trial drops, is usually carried out to determine the couple (M_e, d) satisfying Equation (11) with a given level of approximation (Figure 6a).

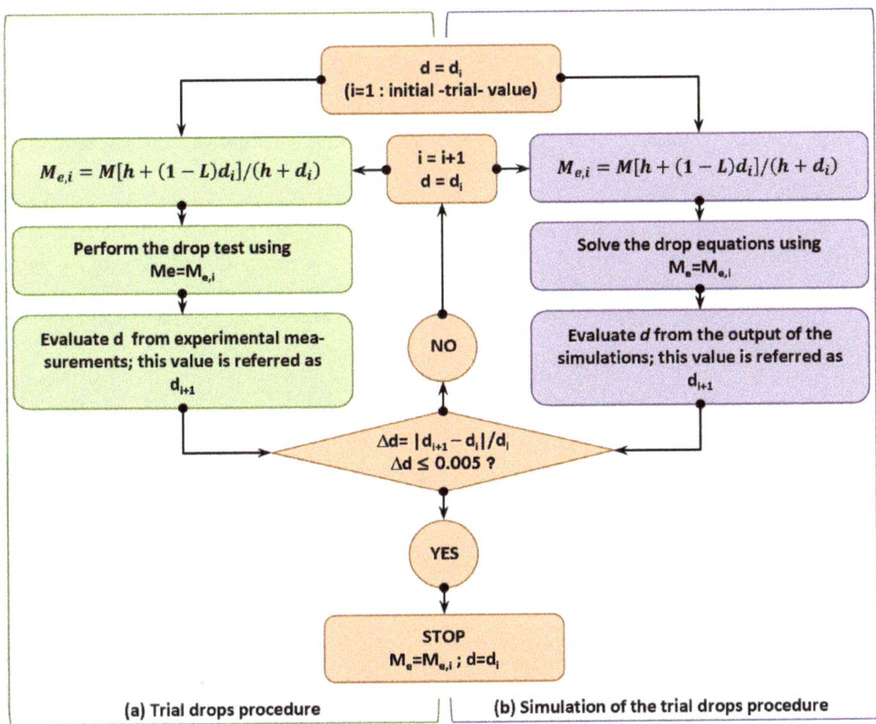

Figure 6. Iterative process for determining effective mass: (**a**) trial drop tests and (**b**) numerical method.

Although the M_e and d values used during the limit drop tests were already available from the landing gear's test reports, they were calculated by simulating the trial drop procedure usually carried out before the limit drop test (Figure 6b); this was clearly done to get a first proof of validation for the proposed numerical formulations.

An initial (trial) value, d_1, was imposed on d, and then, the corresponding effective mass $M_{e,1}$ was obtained by Equation (11); L was imposed as equal to 2/3, and M, as equal to half of the *MLW* ($M = 1520$ kg). Drop dynamics equations were solved, thus giving in output the effective value d_2 of parameter d. A new value of effective mass $M_{e,2}$ was then recalculated by Equation (11), this time corresponding to $d = d_2$; the input parameters $M_{e,2}$, d_2, were used for a new drop simulation, leading to a new output value d_3.

The process was repeated until the difference, Δd, between the values of d obtained in two consecutive iterations were lower than 0.005 m.

After three iterations, the convergence of the process was reached, resulting in $\Delta d = 0.003$ m, $d = 0.360$ m, and $M_e = 1083$ kg. Obtained values for d and M_e were in full agreement with those reported in the drop test documentation of the reference landing gear ($d = 0.45$ m, $M_e = 1090$ kg).

At each iteration, the drop dynamics equations were integrated starting from the initial conditions (listed in Table 1), considering a time step of 5×10^{-4} s over an interval of 1 s.

Table 1. Initial conditions used for integration of drop dynamic equations.

$z_1(0)$	=	$h = 0.475$ m ($\rightarrow t_{imp} = (2\,h/g)^{0.5} = 0.31$ s)
$z_2(0)$	=	$h + L^* + L_1 + L_0 = (0.475 + 0.329 + 0.287 + 0.165) = 1.256$ m
$x_1(t_{imp})$	=	0
$\varphi(0)$	=	0
$\dot{z}_1(0)$	=	0
$\dot{z}_2(0)$	=	0
$\dot{x}_1(t_{imp})$	=	$V_0 = 45.276$ m/s (landing speed, equal to $1.2 \cdot V_{s0}$, where V_{s0} is the aircraft stall speed with flaps down)
$\dot{\varphi}(t_{imp})$	=	$V_0/R = 154.65$ rad/s, where R is the nominal radius of the tire (equal to 0.254 m for the reference gear unit)

The complete set of data related to the reference landing gear and used to run the simulations is reported in Appendix B.

The output related to the last iteration was considered relevant for the drop test under investigation, as it led to coherent values of M_e and d.

The time histories obtained for the variables z_1, z_2, and z_0 are reported in Figure 7.

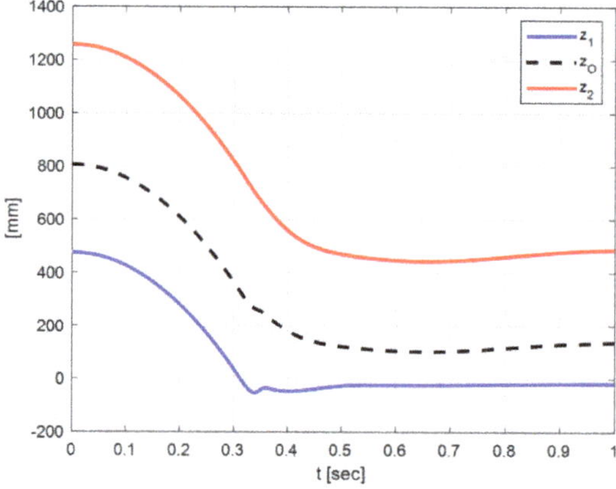

Figure 7. Time histories of vertical displacements z_1, z_0, and z_2.

Before the impact ($t < t_{imp}$), the landing gear and drop mass are in free fall and all the variables show the same parabolic trend versus time. Therefore, the piston stroke is identically null (Figure 8) and no work is done by the shock absorber.

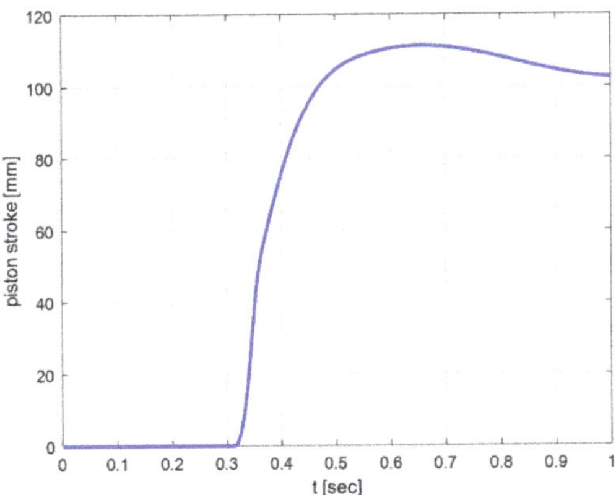

Figure 8. Time history of piston stroke.

After the impact, the piston stroke increases up to its maximum value; in this time interval, z_1, z_0, and z_2 continue to decrease coherently to what is expected during the shock absorber's compression. Their trends are now different and no longer parabolic, as they are influenced by the shock absorber dynamics and the crushing of the tire. The tire's crushing effect is more evident in the wheel hub vertical displacement (z_1) and becomes less dominant for z_0 and z_2 due to the shock absorber's damping action.

When the piston stroke decreases, the shock absorber starts its extension phase and the displacements z_1, z_0, and z_2 consequently increase.

During this phase, the vertical displacement trends are practically affected only by the shock absorber's dissipation force.

This is even more evident if we consider the time histories of the effective mass acceleration ($\ddot{z}_2(t)$) (Figure 9) and the overall shock absorber force $F_{sa}(t) = (F_a(t) + F_d(t))$ (Figure 10).

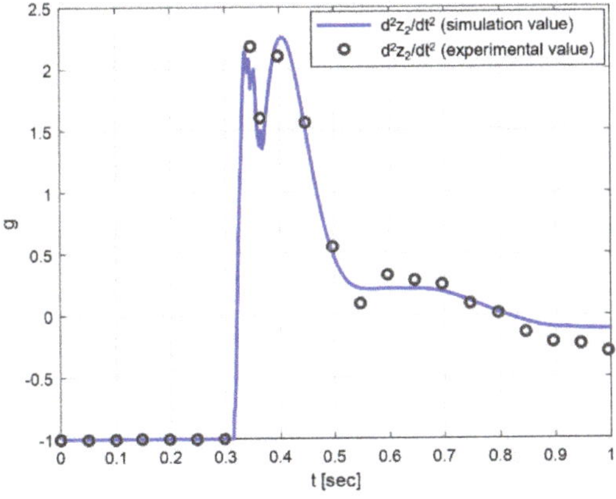

Figure 9. Time history of effective mass acceleration along vertical axis.

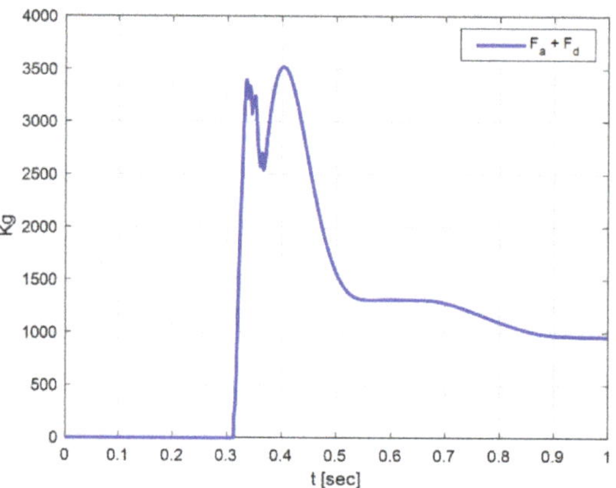

Figure 10. Time history of shock absorber total force ($F_a + F_d$).

Since the effect of F_{PN} on $\ddot{z}_2$ is negligible along the entire piston stroke, the effective mass experiences an enforced motion that is practically driven only by the effective mass weight and overall shock absorber force. Thus, as shown by the results plotted in Figures 9 and 10, the time histories of $\ddot{z}_2(t)$ and $F_{sa}(t)$ have similar trends, represented by the linear relation $M_e \ddot{z}_2(t) \approx F_{sa}(t) - M_e g$.

The acceleration of the effective mass was one of the two parameters measured during the drop test, using a monoaxial accelerometer placed at the top of the effective mass (Figure 11).

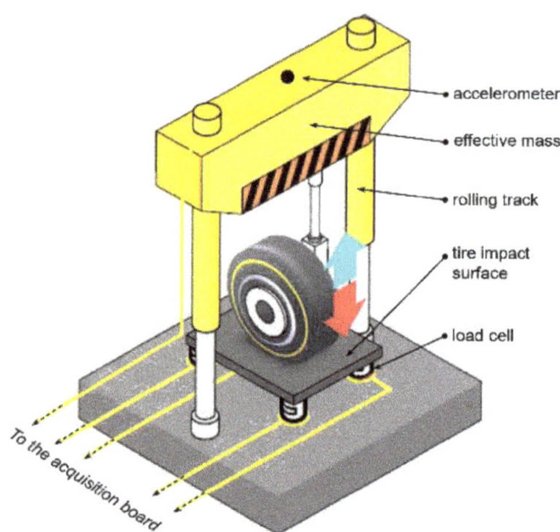

Figure 11. Drop test arrangement (sketch).

The nominal acquisition range of the accelerometer was equal to 3 Hz to 5 KHz; during the test, the vertical acceleration was acquired with a sampling frequency of 100 Hz. The most significant measurement points are reported in Figure 9, and they show excellent agreement between numerical expectations and experimental findings (to enhance the quality of the figure, the experimental data were plotted with a marker every 0.05 s).

The ground reaction force (F_z) was the other parameter measured during the drop test; experimental data were obtained by placing four load cells under the tire's impact area (Figure 11). Each load cell measured along the normal to the plane of impact (*z*-axis) and was characterized by a nominal acquisition range of 0.08 Hz to 25 KHz and an allowable compressive load of 5000 lb; for the acquisition of vertical load, a sampling frequency of 100 Hz was set. The four outputs of the cells were summed to get the experimental trend of F_Z; with the experimental error of each measurement equal to ± 10 lb, the overall experimental uncertainty of F_Z values was equal to ± 40 lb (± 18.14 kg).

By working on simulation output, F_z was numerically evaluated as $-F_{PN}(t)$, and its trend satisfactorily followed the experimental measurements (Figure 12); to enhance the quality of the graph, the experimental data were plotted with a marker every 0.05 s.

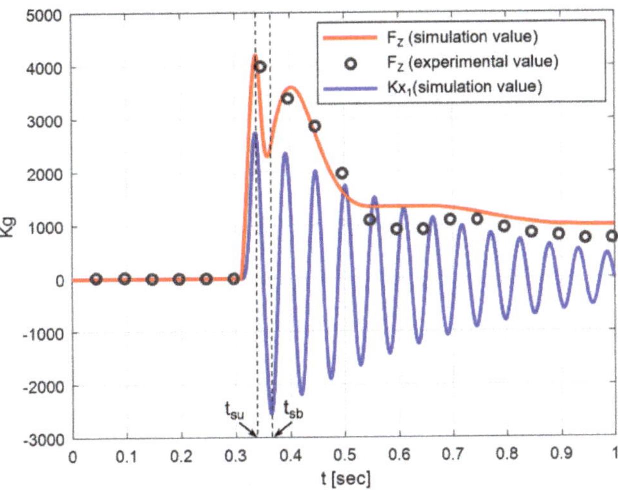

Figure 12. Time history of vertical and horizontal forces at the hub.

The theoretical time history of the horizontal force acting on the wheel hub is reported in Figure 12 to facilitate a discussion on its consistency with $F_z(t)$.

Before the impact, the tire has relative horizontal speed with respect to the ground, which is equal to the aircraft's landing speed.

At touchdown, the horizontal friction force between the tire and the ground induces elastic displacement of the wheel hub along the x-axis (x_1). A recovery force due to the elasticity of the landing gear leg then arises (Kx_1), and the damped oscillatory motion of the wheel hub begins.

The maximum and minimum values of the elastic recovery force are referred to as spin-up and spring back, respectively; as shown in Figure 12, they occur at $t_{su} = 0.33$ s and $t_{sb} = 0.37$ s.

When $t_{su} < t < t_{sb}$, vertical force F_z abruptly decreases; this is in full agreement with the physics of the phenomenon, as the ground impact force is alleviated by the forward bending of the landing gear leg. As also confirmed by the experimental data, after t_{sb} F_z increases because of the amplification induced by the rearward bending of the landing gear leg, and then its trend becomes prevalently influenced by the shock absorber dynamics rather than by the horizontal (elastic) deflections of the leg.

The initial oscillation of F_z vs. time is essentially due to the elastic response of the landing gear along the x-axis; a first peak arises when the horizontal force reaches the spin-up value ($t = t_{su}$), then the vertical load decreases up to a local minimum occurring at $t = t_{sb}$, that is, when the horizontal load reaches the spring-back value. Since the numerical trend of F_z follows the same initial oscillation resulting from the experiment, it is indirectly proven that the landing gear motion along the x-axis is coherently reproduced in terms of instants of time when spin-up and spring-back occur.

Moreover, during this oscillation, the amplitude of $F_z(t)$ is deeply influenced by the elastic bending of the landing gear leg, which is in turn driven by the horizontal load at the wheel; since, at t_{su} and t_{sb}, the amplitude of F_z is well predicted by the model, it is reasonable to deduce that the same applies to the amplitude of the spin-up and spring-back loads (even if no direct measurement was made for these entities during the experiment).

In light of all of these considerations, it can be concluded that the proposed numerical method coherently captured the physics of the drop dynamics along both the vertical and horizontal axis of motion while producing results that are fully consistent with experimental outcomes.

4. Conclusions

A numerical method was developed to simulate the impact dynamics of oleo-pneumatic landing gear characterized by trailing-link or telescopic leg arrangements.

The method uses a rational formulation of the differential equation of motion involving only four state variables for the entire system: vertical (rigid) displacement at the cylinder head and piston base, (rigid) rotation of the wheel, and horizontal (elastic) displacement of the wheel hub.

Despite the limited number of variables, all the relevant forces that arise during the drop were duly simulated, along with their effects on the landing gear's motion along the vertical and horizontal axes. A set of intercommunicating routines was implemented in MATLAB environment to integrate the dynamic impact equations, starting from user-defined initial conditions and general parameters related to geometric and structural landing gear layouts. This tool was then used to simulate a drop test of a reference landing gear, and the obtained results were successfully validated against available experimental data.

Besides the good correlation level between numerical predictions and experimental measurements, the proposed approach proved extremely reliable in capturing the physics of all the complex phenomena occurring at drop-impact, including the spin-up and spring-back of the landing gear leg, commonly simulated only through FE-based methodologies. Owing to this remarkable performance and the low level of sophistication of the adopted models, the proposed numerical approach can be reasonably considered as a fast, powerful, and effective tool for supporting engineering activities that usually take place during the development of landing gear units, from the preliminary design of the shock absorber to the design optimization of its main components, and from the refinement of already consolidated layouts to the virtual testing of mature design solutions before manufacturing the test item.

On the other hand, some improvements are necessary to increase the versatility of the proposed methodology for different possible applications. Focusing on the specific certification tasks the drop-simulation is intended to support, adjustments to the formulation are expected to be made in order to overcome current limitations in the following aspects:

-   Landing gear type: Although the single-leg oleo-pneumatic landing gear represents a very common layout, multi-shock-absorber/multi-wheel arrangements need to be included in the model to properly address drop-test simulations in the framework of certification processes for large airplanes.
-   Landing gear free-fall attitude: The assumption of free fall along the normal to the ground needs to be relaxed in order to enable the simulation of drop-tests onto inclined planes, generally recommended by airworthiness regulations to account for particular or emergency landing conditions.

Funding: Part of the research described in this paper has been carried out in the framework of TIVANO Project, which was funded by the Italian Ministery of Research and Instruction, and the Italian Ministery of the Economic Development, under the National Operative Plan (PON) for research and competitiveness, years 2007–2013; research project code CTN01_00236_256622.

Institutional Review Board Statement: Not applicable.

Informed Consent Statement: Not applicable.

Data Availability Statement: All data used in this study have been declared in the paper's main text and Appendix B.

Acknowledgments: The author would like to thank Leonardo Aircraft S.p.A., for the industrial guidelines and relevant input data supporting the validation of the adopted simulation approaches.

Conflicts of Interest: The author declares no conflict of interest. The funders had no role in the design of the study; in the collection, analyses, or interpretation of data; in the writing of the manuscript, or in the decision to publish the results.

Nomenclature

A_0	air chamber cross section area
A_s	orifice net section area
A_{tr}	portion of the orifice section area crossed by the oil-flow
c	(piston) stroke
c_{visc}	viscosity coefficient of the oil
d	deflection under impact of the tire
dof	degrees of freedom
FEM/FEA	finite element model/finite element analysis
F_a	reaction force exerted by the gas contained in the air-chamber
F_d	dissipation force associated to the oil flow across the orifice
F_{PN}	reaction force due to tire crushing
F_x	friction force between the tire and the ground
F_Z	overall vertical (reaction) force at the tire-to-ground point of contact
g	acceleration of gravity
h	drop height
I	wheel polar moment of inertia
K	stiffness of the landing gear leg along the horizontal (x-) axis
L	lift over weight ratio
L^*	vertical distance between points N and O before the impact
L_0	air-chamber axial length before the impact
L_1	piston length
L_2	vertical distance of the point N to the head of the cylnder
m	swinging mass
M	static weight on the main landing gear, with aircraft in the level attitude and the noose wheel clear
M_e	effective mass
MLW	maximum landing weight
MTOW	maximum take-off weight
P_0	preload pressure
R	tire nominal radius (no crushing)
s	x-displacement of the point of contact between the tire and the ground
SB/sb	spring-back
SU/su	spin-up
t	time-instant
t_{imp}	time-instant when the tire impacts the ground
t_{sb}	time-instant when the spring-back occurs
t_{su}	time-instant when the spin-up occurs
V_0	landing speed (parallel to the ground)
vs	versus
x_1	horizontal displacement of the wheel's hub
z_0	vertical displacement of the point O, placed at the base of the piston
z_1	vertical displacement of the wheel's hub
z_2	vertical displacement of the cylinder head
Greek symbols	
γ	polytropic index
φ	wheel's rotation angle

μ	sliding (or kinetic) friction coefficient
ρ	radius of the wheel (after tire crushing)
ρ_{oil}	oil density
σ	viscous structural damping
θ	cylinder axis inclination with respect to the normal to the ground

Appendix A. Relation among the Piston Stroke and the Lagrangian Variables z_1 and z_2 in Case of Landing Gear with Trailing Link

As seen in Section 2, the expression of the piston stroke for landing gears with trailing links is given by:

$$c(t) = z_O(t) - z_2(t) + L_1 + L_0$$

where:

- L_0 and L_1 are the longitudinal lengths of the air-chamber and piston, respectively;
- $z_2(t)$ is the Lagrangian coordinate representing the vertical displacement of the cylinder head;
- $z_0(t)$ is the vertical displacement of the point O, located at the base of the piston at the center of the piston-fork hinged joint.

$z_0(t)$ does not represent a different state variable of the system, as it is dependent on both z_2 and z_1, the latter being the vertical displacement of the hub.

The relation among z_0, z_1, and z_2 will be here derived from a detailed analysis of trailing link mechanics.

Referring to the scheme of Figure A1, let M be the point at the center of the hub axis, and N the point at the center of the hinged connection between the fork and the trailing link.

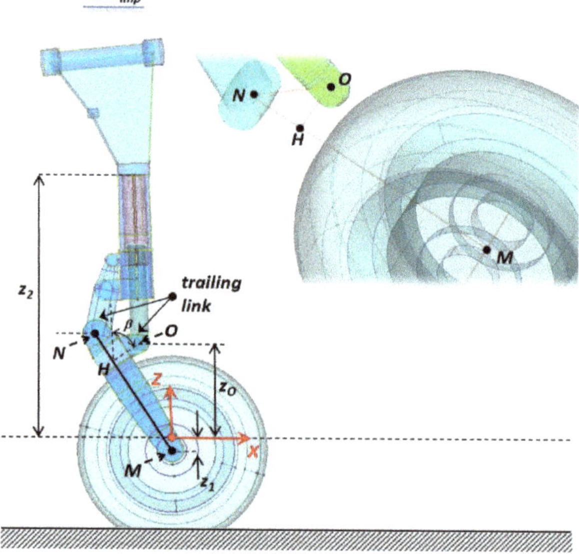

Figure A1. Landing gear with trailing link after the impact.

The vertical coordinate, z, of the generic point on segment $\overline{MN}$ located at a distance ε from M can be expressed as:

$$z(\varepsilon) = \frac{MN - \varepsilon}{MN} z_N + \frac{\epsilon}{MN} z_M \tag{A1}$$

where z_M and z_N are the z-coordinates of M and N, while MN is the length of the segment $\overline{MN}$.

If H indicates the projection of O on $\overline{MN}$, then, it results at any instant of time:

$$z_O = z_H + OH cos\beta \tag{A2}$$

with β denoting the angle between the segment $\overline{OH}$ and the normal to the ground passing through point H.

According to Equation (A1), the z-coordinate of the point H is equal to:

$$z_H = z(\varepsilon)|_{\varepsilon=MH} = \frac{MN - MH}{MN}z_N + \frac{MH}{MN}z_M$$

and, therefore, by substitution into Equation (A2), it can be obtained for z_0:

$$z_O = \frac{MN - MH}{MN}z_N + \frac{MH}{MN}z_M + OH cos\beta. \tag{A3}$$

Let N^*, H^*, and O^* indicate the positions of the points N, H, and O at the instant of the impact; referring to the scheme of Figure A2, it can be observed that, at the generic instant of time $t > t_{inp}$, the following applies:

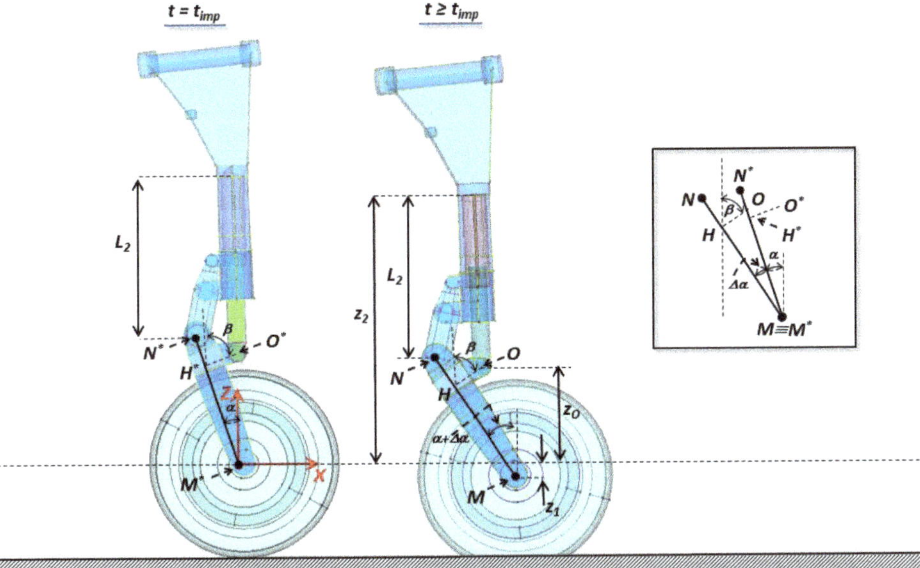

Figure A2. Reference scheme for the evaluation of the angle β.

$$\beta = \frac{\pi}{2} - (\alpha + \Delta\alpha) \ \rightarrow \ cos\beta = sin(\alpha + \Delta\alpha) \tag{A4}$$

$\Delta\alpha$ being the angle swept by the segment $\overline{MN}$ while rotating around M during the time interval $[t_{inp}: t]$.

Recalling some trigonometrical identities, $sin(\alpha + \Delta\alpha)$ may be rearranged as:

$$sin(\alpha + \Delta\alpha) = \sqrt{1 - cos^2(\alpha + \Delta\alpha)} = \sqrt{1 - \left(\frac{z_N - z_M}{MN}\right)^2}. \tag{A5}$$

By taking in account Equations (A4) and (A5), Equation (A3) turns into:

$$z_O = \frac{MN - MH}{MN} z_N + \frac{MH}{MN} z_M + OH \sqrt{1 - \left(\frac{z_N - z_M}{MN} \right)^2} \tag{A6}$$

and, finally, after the substitutions $z_M = z_1$; $z_N = z_2 - L_2$ (see Figure A2):

$$z_O = z_O(z_1, z_2) = \frac{MN - MH}{MN}(z_2 - L_2) + \frac{MH}{MN} z_1 + OH \sqrt{1 - \left(\frac{z_2 - L_2 - z_1}{MN} \right)^2} \tag{A7}$$

Appendix B

The entire set of input data feeding the simulations addressed by this paper have been collected in the following tables.

The interested reader may refer to this appendix to get all the necessary information to reproduce and verify the proposed numerical method and related formulations.

Table A1. Reference landing gear input data: geometric characteristics, inertial properties and stiffness along the x-axis

Parameter	Value
L_0	0.165 m
L_1	0.287 m
L_2	0.401 m
L^*	0.330 m
R	0.254 m
MN	0.403 m
MH	0.317 m
NH	0.087 m
OH	0.092 m
I	0.52 Kg m^2
m	36.84
K	6.3287×10^5 N/m

Table A2. Tire mechanical properties (F_{PN} vs. tire crushing at nominal inflation pressure).

Tire Crushing (m)	F_{PN} (N)
0	0.00
0.001	390.12
0.002	796.10
0.003	1217.92
0.004	1655.60
0.005	2109.12
0.006	2578.49
0.007	3063.71
0.008	3564.78
0.009	4081.70
0.01	4614.47
0.02	10,813.88
0.03	18,598.23
0.04	27,967.52
0.05	38,921.75
0.06	51,460.92
0.07	65,585.03
0.08	73,241.44

Table A3. Shock absorber main characteristics.

Parameter	Value
A_0	1.77×10^{-3} m^2
A_S	3.15×10^{4} m^2
P_0	1.17×10^{6} Pa
c_{visc}	0.70 N·s/m
ρ_{oil}	920 Kg/m^3

Table A4. Dissipative force coefficient and pneumatic reaction vs. piston stroke.

Piston Stroke [m]	$F_d/\dot{c}^2$ [Ns^2/m] *	F_a [N]
0.000	5.96×10^5	5.52×10^{-6}
0.001	5.85×10^5	5.57×10^{-6}
0.004	5.35×10^5	5.83×10^{-6}
0.008	2.89×10^5	7.93×10^{-6}
0.009	2.57×10^5	8.41×10^{-6}
0.011	2.00×10^5	9.54×10^{-6}
0.014	1.59×10^5	1.07×10^{-5}
0.019	1.05×10^5	1.31×10^{-5}
0.021	8.07×10^4	1.50×10^{-5}
0.024	6.74×10^4	1.64×10^{-5}
0.026	5.52×10^4	1.81×10^{-5}
0.036	3.68×10^4	2.22×10^{-5}
0.041	3.68×10^4	2.22×10^{-5}
0.046	3.95×10^4	2.15×10^{-5}
0.051	5.29×10^4	1.85×10^{-5}
0.056	7.01×10^4	1.61×10^{-5}
0.059	8.39×10^4	1.47×10^{-5}
0.061	1.01×10^5	1.34×10^{-5}
0.064	1.43×10^5	1.13×10^{-5}
0.066	1.50×10^5	1.10×10^{-5}
0.069	1.55×10^5	1.08×10^{-5}
0.071	1.63×10^5	1.05×10^{-5}
0.081	1.95×10^5	9.66×10^{-6}
0.091	2.36×10^5	8.77×10^{-6}
0.101	2.80×10^5	8.05×10^{-6}
0.111	3.53×10^5	7.17×10^{-6}
0.121	4.54×10^5	6.32×10^{-6}
0.133	6.44×10^5	5.31×10^{-6}

* For each value in this column, the corresponding A_{tr} may be obtained from Equation (4) Section 2.

Table A5. Drop simulation settings (relevant data).

Parameter	Value
h	0.475 m
V_0	45.28 m/s
M	1520 Kg
M_e	1083 Kg
σ	0.02
μ	0.75

References

1. Young, D.W. Aircraft Landing Gears—The Past, Present and Future. *Proc. Inst. Mech. Eng. Part D J. Automob. Eng.* **1986**, *200*, 75–92. [CrossRef]
2. *EASA Part 66—Module 11.13—Landing Gear*; European Union Aviation Safety Agency: Köln, Germany, 2015.

3. *FAA-H-8083-31A, Aviation Maintenance Technician Handbook—Airframe Vol. 2, U.S.*; Department of Transportation, Federal Aviation Administration, Flight Standards Service: Washington, DC, USA, 2018.
4. Hadekel, R. Shock absorber calculations. *Aircr. Eng.* **1940**, *19*, 71–73.
5. Milwitzky, B.; Cook, F.E. *Analysis of Landing-Gear Behavior*; NACA Report 1154; Langley Field: Hampton, VA, USA, 1953; p. 45.
6. Currey, N.S. *Aircraft Landing Gear Design: Principles and Practices*; AIAA Inc.: Marietta, GA, USA, 1988.
7. Daniels, J.N. *A Method for Landing Gear Modeling and Simulation with Experimental Data*; NASA Contractor Report 201601; National Aeronautics and Space Administration, Langley Research Center: Hampton, VA, USA, 1996.
8. Sivakumar, S.; Haran, A.P. Mathematical model and vibration analysis of aircraft with active landing gears. *J. Vib. Control.* **2015**, *21*, 229–245. [CrossRef]
9. Sivakumar, S.; Haran, A.P. Aircraft random vibration analysis using active landing gears. *J. Low Freq. Noise Vib. Active Control* **2015**, *34*, 307–322. [CrossRef]
10. Sivakumar, S.; Syedhaleem, M. Nonlinear vibration analysis of oleo pneumatic landing gear at touchdown impact. *Math. Models Eng.* **2018**, *4*, 89–97. [CrossRef]
11. ABAQUS Software General Description Webpage. Available online: https://www.3ds.com/ (accessed on 26 February 2021).
12. Park, I.K.; Kim, S.J.; Ahn, S.M. Spin-up, Spring-back Load Analysis of KC-100 Nose Landing Gear using Explicit Finite Element Method. *J. Korean Soc. Aviat. Aeronaut.* **2011**, *19*, 51–57.
13. Kruger, W.R.; Moroandini, M. Numerical Simulation of Landing Gear Dynamics: State-of-the-art and Recent Developments. In Proceedings of the AVT-152 Symposium on Limit Cycle Oscillation and other Amplitude-Limited Self-Excited Vibrations, Loen, Norway, 5–8 May 2008.
14. Lyle, K.H.; Jackson, K.E.; Fasanella, E.L. Simulation of aircraft landing gears with a nonlinear dynamic finite element code. *J. Aircr.* **2002**, *39*, 142–147. [CrossRef]
15. Khapane, P.D. Gear walk instability studies using flexible multi-body dynamics simulation methods in Simpack. *Aerosp. Sci. Technol.* **2006**, *10*, 19–25. [CrossRef]
16. Lernbeiss, R.; Plöch, M. Simulation model of an aircraft landing gear considering elastic properties of the shock absorber. *J. Multi-Body Dyn.* **2007**, *221*, 78–86. [CrossRef]
17. Kong, J.P.; Lee, Y.S.; Han, J.D.; Ahn, O.S. Drop impact analysis of smart unmanned aerial vehicle (SUAV) landing gear and comparison with experimental data. *Mater. Sci. Eng. Technol.* **2009**, *40*, 192–197. [CrossRef]
18. Xue, C.J.; Han, Y.; Qi, W.G.; Dai, J.H. Landing-Gear drop-test rig development and application for light airplanes. *J. Aircr.* **2012**, *49*, 2064–2076. [CrossRef]
19. MATLAB Software General Description Webpage. Available online: https://www.mathworks.com/products/matlab.html (accessed on 26 February 2021).
20. Carnahan, B.; Luther, H.A.; Wilkes, J.O. *Applied Numerical Methods*; John Wiley and Sons Inc.: New York, NY, USA, 1969; Chapter 6.
21. *STANAG 4671—Unmanned Aerial Vehicles Systems Airworthiness Requirements*, 1st ed.; NATO Standardization Agency: Brussels, Belgium, 2009.

applied sciences

Article

Design and Validation of Disturbance Rejection Dynamic Inverse Control for a Tailless Aircraft in Wind Tunnel

Bowen Nie [1,2,*], Zhitao Liu [2], Tianhao Guo [2,*], Litao Fan [2], Hongxu Ma [1] and Olivier Sename [3]

[1] College of Intelligence Science and Technology, National University of Defense Technology, Changsha 410073, China; mhx1966@163.com

[2] Low Speed Aerodynamics Institute, China Aerodynamics Research and Development Center, Mianyang 621000, China; liuzhitao@cardc.cn (Z.L.); litaofan_cardc@outlook.com (L.F.)

[3] Grenoble INP (Institute of Engineering, Université Grenoble-Alpes), CNRS, GIPSA-Lab, 38000 Grenoble, France; olivier.sename@gipsa-lab.grenoble-inp.fr

* Correspondence: niebowen_cardc@outlook.com (B.N.); guotianhao@nudt.edu.cn (T.G.)

Abstract: This paper focuses on the design of a disturbance rejection controller for a tailless aircraft based on the technique of nonlinear dynamic inversion (NDI). The tailless aircraft model mounted on a three degree-of-freedom (3-DOF) dynamic rig in the wind tunnel is modeled as a nonlinear affine system subject to mismatched disturbances. First of all, a baseline NDI attitude controller is designed for sufficient stability and good reference tracking performance of the nominal system. Then, a nonlinear disturbance observer (NDO) is supplemented to the baseline NDI controller to estimate the lumped disturbances for compensation, including unmodeled dynamics, parameter uncertainties, and external disturbances. Mathematical analysis demonstrates the convergence of the employed NDO and the resulting closed-loop system. Furthermore, an anti-windup modification is applied to the NDO for control performance preserving in the presence of actuator saturation. Subsequently, the designed control schemes are preliminarily validated and compared via simulations. The baseline NDI controller demonstrates satisfactory attitude tracking performance in the case of nominal simulation; the NDO augmented NDI controller presents significantly improved ability of disturbance rejection when compared with the baseline NDI controller in the case of robust simulation; the anti-windup modified scheme, rather than the baseline NDI controller nor the NDO augmented NDI controller, can preserve the closed-loop performance in the case of actuator saturation. Finally, the baseline NDI scheme and the NDO augmented NDI scheme are implemented and further validated in the wind tunnel flight tests, which demonstrate that the experimental results are in good agreement with that of the simulations.

Keywords: tailless aircraft; nonlinear dynamic inversion; disturbance observer; anti-windup; flight tests; wind tunnel

Citation: Nie, B.; Liu, Z.; Guo, T.; Fan, L.; Ma, H.; Sename, O. Design and Validation of Disturbance Rejection Dynamic Inverse Control for a Tailless Aircraft in Wind Tunnel. *Appl. Sci.* **2021**, *11*, 1407. https://doi.org/10.3390/app11041407

Academic Editor: Ruxandra Mihaela Botez

Received: 29 December 2020
Accepted: 1 February 2021
Published: 4 February 2021

1. Introduction

Tailless aircraft configurations have gained considerable attention due to the inherent increase in stealth and decreases in weight and drag [1,2]. However, flight control law design for the tailless aircraft is challenging due to the multiaxis instabilities, the insufficient yaw control power, the inaccurate control-oriented modeling, and the adverse external disturbances [1,3,4]. Critical flight control research problems for tailless aircraft exist in yaw departure and recovery [5,6], multivariable control for good flying and handling qualities [7–12], reconfigurable control for failure or damage tolerance [13–17], thrust vectoring for envelope expansion [18,19], etc.

In previous aviation industry practices, nonlinear dynamic inversion (NDI) has played an important role in the field of tailless aircraft control law design [3,8,9,20], which provides a nice compromise between controller complexity and performance. NDI is a nonlinear control approach that cancels the system nonlinearity using an onboard dynamic model and feedback. The resulting dynamics of selected control variables are globally reduced to integrators across the operating regimes. Hence, simple controllers without gain schedule are adequate to regulate the control variables for desirable closed-loop dynamics [21]. In principle, a perfect knowledge of the system dynamics across the entire flight envelope is required to achieve an exact dynamic cancellation. However, such a requirement is almost impossible to meet in reality due to modeling simplifications, computational errors, and external disturbances [22,23]. Particularly, for the tailless aircraft, uncertainties and disturbances come from the following possible sources: aerodynamic and propulsive approximations, neglected control effector interactions, neglected vehicle elasticity, unmodeled actuator and sensor dynamics, time delay in the feedback path, and wind gust [9]. Besides, there are higher levels of model errors in case of failures, damage, or highly nonlinear phenomena [24]. Intuitively, to overcome the potential robustness issues, flight control researchers employ μ synthesize, adaptive neural network to construct a robust outer loop for the NDI controller of the tailless aircrafts such as the ICE (innovative control effector) [8,9] and the X-36 [16,17]. Although the robustness of these techniques is superior to the regular NDI, not all the uncertainties nor the control signal constraints are taken into account [25], which are inevitably present in the real flight context of tailless aircraft. Meanwhile, to provide sufficient control effort to suppress the effects of the disturbances, such robust methods tend to have large feedback gains [26]. Consequently, a possible closed-loop performance reduction is caused by the conservative design of the gain factors [27].

An alternative solution is based on a composite control structure, including a baseline NDI controller for the desired performance specifications of the nominal system and a nonlinear disturbance observer (NDO) for robustness enhancement of the closed-loop system. The fundamental idea of NDO is to bring together all the internal uncertainties, external disturbances, parameter variations, and unmodeled dynamics as lumped disturbances [26], which are always mismatched and cannot be completely counteracted through control input channels [28]. Taking the partially known nonlinear dynamics into account, an appropriately designed NDO can be employed to reconstruct an inverse model of the actual system, and estimate the nonlinear lumped disturbances by using the baseline NDI control input and the system output. As described in ref [25,28], it is feasible to remove the influence of lumped disturbances from the system output variables in the steady-state and reduce the influence on transient performance with a composite controller involving the NDI and NDO. The NDO employed in this paper is firstly proposed by Chen with rigorous stability analysis of both the NDO and the closed-loop system [29], and then widely applied to the field of flight control, such as the longitudinal autopilot of a missile [30], the longitudinal tracking of an air-breathing hypersonic vehicle [27], the attitude control of a spacecraft [31], etc. It appears that such NDO based control methods can lead to significant improvement of the disturbance rejection ability of an existing controller without scarifying the nominal performance. However, to the authors' knowledge, the existing literatures on NDO based NDI (NDI-DO) control of tailless aircraft are rare. Though a primitive NDI-DO controller has been designed for a tailless aircraft and numerically validated in ref. [32], experimental implementation and flight validation are to be carried out.

As is well known, another main weakness of NDI is the requirement of the plant's input distribution matrix to be invertible. However, it is often the case that the input distribution matrix becomes "almost" singular for some states and hence may lead to excessively large control activities [33]. As deflection range and rate limitations are inevitably present in physical actuated surfaces, aggressively piloted maneuvers of aircraft may yield actuator saturation and associated degradation of closed-loop performance or even loss of stability [34]. A favorable solution in industry and academia to address the actuator saturation is the so-called anti-windup (AW) techniques, in which an existing controller is augmented with an additional element that only becomes active during saturation [35]. In ref. [25], a classic static AW scheme is added to the baseline controller of a tailless unmanned aerial vehicle (UAV) allowing for windup attenuation and nominal performance recovery in the absence of saturation. This paper extends the work of ref. [25] to the NDI setting and nonlinear case.

The remainder of this paper is structured as follows. Section 2 provides the preliminaries including a brief introduction to the tailless aircraft, the experimental setup and the mathematical modeling. In Section 3, the equations of angular motions are reformulated in the form of a nonlinear affine system for application of the NDI control scheme, which is then augmented with an NDO to improve the ability of disturbance rejection and further modified for anti-windup enhancement in the presence of actuator saturation. In Section 4, the designed baseline NDI controller, NDI-DO controller, and anti-windup modified controller (NDI-AW) are validated and compared in the context of nonlinear simulations. In Section 5, both the baseline NDI scheme and the NDI-DO scheme are implemented for free flight tests in the wind tunnel, in which the NDI-DO scheme demonstrates the superior ability of disturbance rejection in contrast with the baseline NDI scheme. Finally, a conclusion is provided in Section 6.

2. Preliminaries

2.1. Tailless Aircraft Experimental Setup

The studied tailless aircraft is of the medium-aspect-ratio and blended-wing-body configuration as shown in Figure 1. There are eight individually driven control surfaces, which could realize the traditionally defined control deflections, i.e., the elevator, the aileron and the rudder. The elevators are classified in left and right having positive deflection when turning downward, the ailerons are also classified left and right having positive deflection when the left surface goes up and the right one goes down. The split drag rudders (SDRs), each one positioned on one side of the wing, are composed of two deferential surfaces: a trailing-wing flap and a mid-wing spoiler. The positive deflection of the rudder is achieved when the left SDR splits symmetrically and the right one keeps at zero.

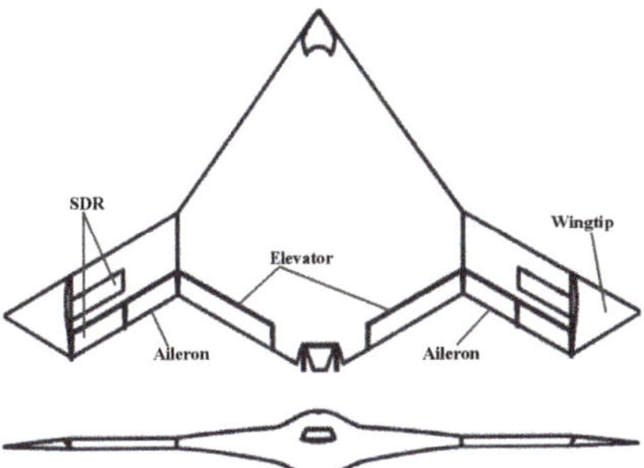

Figure 1. Layout of the tailless aircraft.

From a flight test point of view, the wind tunnel model should be geometrically and dynamically similar to the full-scale aircraft. Hence, a 10% scaled model was fabricated with carbon fiber and aluminum alloy. The physical parameters, including the characteristic size, mass, and inertia, are carefully tuned as listed in Table 1. A spherical joint is selected to connect a vertical support rig to the airframe at the center of gravity (CG), which frees the aircraft's rotational motions (roll/pitch: $\pm 40°$, yaw: $\pm 180°$) and eliminates the translational motions, as shown in Figure 2. The physical parameters of the scaled model are listed in Table 1. The model is equipped with not only the surface actuators, but also the sensors including two vanes for angle of attack (α) and angle of sideslip (β), an inertial measurement unit (IMU) for angular rates (p, q, r), and an attitude and heading reference system (AHRS) for Euler angles (ϕ, θ, ψ). A flight computer is employed for the implementation of control laws and data acquisition (100 Hz) using the technique of rapid control system prototyping. For simplicity, the flight computer is placed outside the wind tunnel test section and wired to the onboard avionics. A "pilot" is responsible for handling the aircraft model with a stick.

Table 1. Physical properties of the dynamically scaled aircraft model.

Name	Symbol	Value
Scale ratio	K_L	10%
Span	b	2.0 m
Mean chord	$\bar{c}$	0.6571 m
Wing area	S	0.8742 m^2
Mass	m	16.85 kg
Moment of inertia	J_x, J_y, J_z	1.1017, 0.7380, 1.6920 kg·m^2

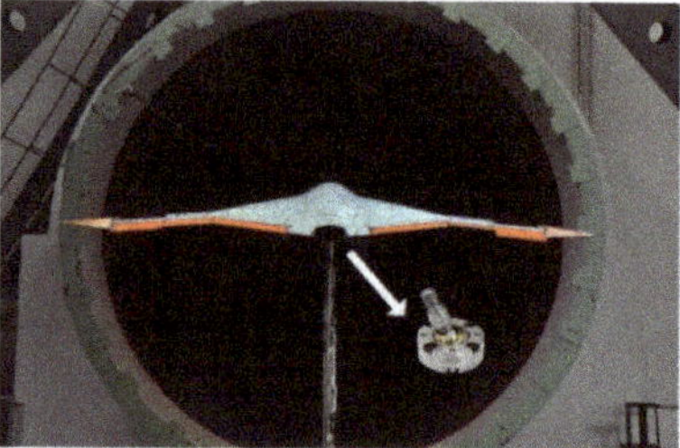

Figure 2. Tailless aircraft model flying on a three degree-of-freedom (3-DOF) dynamic rig in the wind tunnel.

2.2. Mathematical Modeling

As the airflow in the wind tunnel test section is exactly horizontal, the experimental setup demonstrates "1g" level flights, i.e., the flight path angle (γ) is always fixed at zero. The flight dynamics of the tailless aircraft model on the rig are modeled as a three degree-of-freedom (3-DOF) angular motion, which has been developed in ref. [32,36,37]. In modeling of this motion, a set of assumptions are made including: the rigid-body aircraft model, the symmetrical distribution of aircraft mass about the xz plane, the negligible aerodynamic interference induced by the rig and the limited dynamic influence of rig friction and support misalignment.

The equations that describe the angular dynamics are derived from Euler's laws in the body-fixed frame as commonly defined for the aircraft [38].

$$\begin{bmatrix} \dot{p} \\ \dot{q} \\ \dot{r} \end{bmatrix} = J^{-1} M_B - J^{-1} \left(\begin{bmatrix} p \\ q \\ r \end{bmatrix} \times J \begin{bmatrix} p \\ q \\ r \end{bmatrix} \right) \tag{1}$$

where $J = \begin{bmatrix} J_x & 0 & J_{xz} \\ 0 & J_y & 0 \\ J_{xz} & 0 & J_z \end{bmatrix}$.

The CG of the tailless aircraft model coincides well with the aerodynamic reference point. The total external moment acting on the aircraft model M_B is the sum of the aerodynamic moments M_A, the rig friction torque M_F, and the support misalignment induced torque M_G.

$$M_B = M_A + M_F + M_G \tag{2}$$

The kinematic equations of the angular motion are expressed with the aerodynamic angles α, β, and the bank angle about the velocity vector (μ) that defines the rotation between the body and wind frame.

$$\begin{bmatrix} \dot{\alpha} \\ \dot{\beta} \\ \dot{\mu} \end{bmatrix} = \begin{bmatrix} -\tan\beta\cos\alpha & 1 & -\tan\beta\sin\alpha \\ \sin\alpha & 0 & -\cos\alpha \\ \sec\beta\cos\alpha & 0 & \sec\beta\sin\alpha \end{bmatrix} \begin{bmatrix} p \\ q \\ r \end{bmatrix} + \begin{bmatrix} d_{21} \\ d_{22} \\ d_{23} \end{bmatrix} \tag{3}$$

where

$$\begin{bmatrix} d_{21} \\ d_{22} \\ d_{23} \end{bmatrix} = \frac{1}{MV}\left(\begin{bmatrix} W\cos\gamma\cos\mu\sec\beta \\ W\cos\gamma\sin\mu \\ -W\cos\gamma\cos\mu\tan\beta \end{bmatrix}\right.$$
$$\left.+\begin{bmatrix} -L\sec\beta \\ Y\cos\beta \\ L(\tan\beta+\tan\gamma\sin\mu)+Y\tan\gamma\cos\mu\cos\beta \end{bmatrix}\right)$$

It is noted that the last terms in Equation (3), i.e., $\begin{bmatrix} d_{21} & d_{22} & d_{23} \end{bmatrix}^T$ are the translational terms (aerodynamic lift L, aerodynamic side-force Y, weight W, aircraft mass M, and airspeed V) that influence the rotational motion [39]. As widely accepted [40–42], the influence is negligible when designing the attitude controller.

The aerodynamic forces and moments are expressed as:

$$F_A = \begin{bmatrix} L \\ D \\ Y \end{bmatrix} = \frac{1}{2}\rho V^2 S \begin{bmatrix} C_L \\ C_D \\ C_Y \end{bmatrix} \tag{4}$$

$$M_A = \begin{bmatrix} l \\ m \\ n \end{bmatrix} = \frac{1}{2}\rho V^2 S \begin{bmatrix} b\cdot C_l \\ \bar{c}\cdot C_m \\ b\cdot C_n \end{bmatrix} \tag{5}$$

where D is the aerodynamic drag; l, m, n represent the aerodynamic moment of roll, pitch, and yaw; ρ is the air density; $S, b, \bar{c}$ denote the reference wing area, wingspan and mean aerodynamic chord respectively. In practice, the aerodynamic force $\begin{bmatrix} C_L & C_D & C_Y \end{bmatrix}^T$ and moment $\begin{bmatrix} C_l & C_m & C_n \end{bmatrix}^T$ coefficients can be approximated and formulated as the following form [32,37]:

$$C_L = C_{L0}(\alpha,\beta) + \Delta C_{L_{\delta_e}}(\alpha)\delta_e + \Delta C_{L_{\delta_a}}(\alpha,\beta)\delta_a$$
$$+\Delta C_{L_{\delta_r}}(\alpha,\beta)\delta_r + d_{C_L} \tag{6}$$

$$C_D = C_{D0}(\alpha,\beta) + \Delta C_{D_{\delta_e}}(\alpha)\delta_e + \Delta C_{D_{\delta_a}}(\alpha,\beta)\delta_a$$
$$+\Delta C_{D_{\delta_r}}(\alpha,\beta)\delta_r + d_{C_D} \tag{7}$$

$$C_Y = C_{Y0}(\alpha,\beta) + \Delta C_{Y_{\delta_a}}(\alpha,\beta)\delta_a + \Delta C_{Y_{\delta_r}}(\alpha,\beta)\delta_r$$
$$+d_{C_Y} \tag{8}$$

$$C_l = C_{l0}(\alpha,\beta) + \Delta C_{l_{\delta_a}}(\alpha,\beta)\delta_a + \Delta C_{l_{\delta_r}}(\alpha,\beta)\delta_r$$
$$+\hat{C}_{l_p}(\alpha)\hat{p} + \hat{C}_{l_r}(\alpha)\hat{r} + d_{C_l} \tag{9}$$

$$C_m = C_{m0}(\alpha) + \Delta C_{m_{\delta_e}}(\alpha)\delta_e + \Delta C_{m_{\delta_a}}(\alpha,\beta)\delta_a$$
$$+\hat{C}_{m_q}(\alpha)\hat{q} + d_{C_m} \tag{10}$$

$$C_n = C_{n0}(\alpha,\beta) + \Delta C_{n_{\delta_a}}(\alpha,\beta)\delta_a + \Delta C_{n_{\delta_r}}(\alpha,\beta)\delta_r$$
$$+\hat{C}_{n_p}(\alpha)\hat{p} + \hat{C}_{n_r}(\alpha)\hat{r} + d_{C_n} \tag{11}$$

where $\delta_e, \delta_a, \delta_r$ represent the deflection of elevator, aileron and rudder respectively; the normalized body rates are defined as $\hat{p} = pb/2V$, $\hat{q} = q\bar{c}/2V$, and $\hat{r} = rb/2V$. Note that the tailless aircraft takes the SDR as the rudder, which is introduced in Section 2.1.

One can see that the aerodynamic force and moment coefficients are described as a function of the aerodynamic angles, the surface deflections, angular rates, and airspeed. The baseline components of C_{i0}, $i = L, D, Y, l, m, n$ and the incremental components of the surface effectiveness ΔC_{i_j}, $j = \delta_e, \delta_a, \delta_r$ are obtained in the static wind-tunnel tests [36]. The dynamic derivatives of $\hat{C}_{i_k}$, $k = p, q, r$ are measured via the forced-oscillation tests in the wind tunnel [36]. The error components of d_{C_i} are the sums of aerodynamic simplifications and test errors. The resulting aerodynamic database are rearranged and stored in the form of look-up tables.

3. Disturbance Rejection Dynamic Inverse Control

In this section, the NDI control method will be applied to design an attitude controller for the tailless aircraft model in the presence of mismatched disturbances.

3.1. Problem Formulation and Control Objective

For NDI controller design, the equations of angular motions are rewritten in the form of a nonlinear affine system:

$$\begin{cases} \dot{x}_1 = f(x_1, x_2) + g(x_1, x_2)u + d_1 \\ \dot{x}_2 = h(x_1, x_2) + d_2 \end{cases} \tag{12}$$
$$y = x_2$$

with the fast states $x_1 = [p, q, r]^T$, the slow states $x_2 = [\alpha, \beta, \mu]^T$, the input vector $u = [\delta_a, \delta_e, \delta_r]^T$, the output vector y, and the mismatched disturbances of d_1, d_2. The nonlinear behavior of the system lies in the continuous vector functions of $f(x_1, x_2)$, $g(x_1, x_2)$ and $h(x_1, x_2)$.

$$f = J^{-1}\left(-x_1 \times Jx_1 + \begin{bmatrix} QSb\left(C_{l0} + +\hat{C}_{l_p}\hat{p} + \hat{C}_{l_r}\hat{r}\right) \\ QS\bar{c}\left(C_{m0} + \hat{C}_{m_q}\hat{q}\right) \\ QSb\left(C_{n0} + \hat{C}_{n_p}\hat{p} + \hat{C}_{n_r}\hat{r}\right) \end{bmatrix} \right) \tag{13}$$

$$g = J^{-1} \begin{bmatrix} QSb\Delta C_{l_{\delta_a}} & 0 & QSb\Delta C_{l_{\delta_r}} \\ QS\bar{c}\Delta C_{m_{\delta_a}} & QS\bar{c}\Delta C_{m_{\delta_e}} & 0 \\ QSb\Delta C_{n_{\delta_a}} & 0 & QSb\Delta C_{n_{\delta_r}} \end{bmatrix} \tag{14}$$

$$h = \begin{bmatrix} -\tan\beta\cos\alpha & 1 & -\tan\beta\sin\alpha \\ \sin\alpha & 0 & -\cos\alpha \\ \sec\beta\cos\alpha & 0 & \sec\beta\sin\alpha \end{bmatrix} \begin{bmatrix} p \\ q \\ r \end{bmatrix} \tag{15}$$

where $Q = \frac{1}{2}\rho V^2$ is the dynamic pressure.

Obviously, the matrix g is of rank 3 and, therefore, is invertible; the mismatched disturbance d_1 consists of the aerodynamic modeling errors as presented in Equations (9)–(11) and the rig induced unmodeled errors as shown in Equation (2); the mismatched disturbance d_2 mainly results from the neglected dynamics as presented in Equation (3).

In the presence of mismatched disturbances, the control objective is to maintain the attitude of the tailless aircraft model around the trim condition, while also allowing for tracking of a reference command for the states x_2. As a result, the controller is required to provide not only sufficient stability, and good tracking performance, but also rapid disturbance rejection.

3.2. Baseline Dynamic Inversion Control Design

In the absence of the disturbance d_1, d_2, a baseline NDI attitude controller is presented in Figure 3. The fundamental behind the controller configuration is the timescale separation principle. Namely, it is assumed that the attitude dynamics are 10 times faster than the attitude kinematics [43–45]. Consequently, the NDI control law is separated into two control loops: the inner loop for the fast states p, q, r and the outer loop for the slow states α, β, μ.

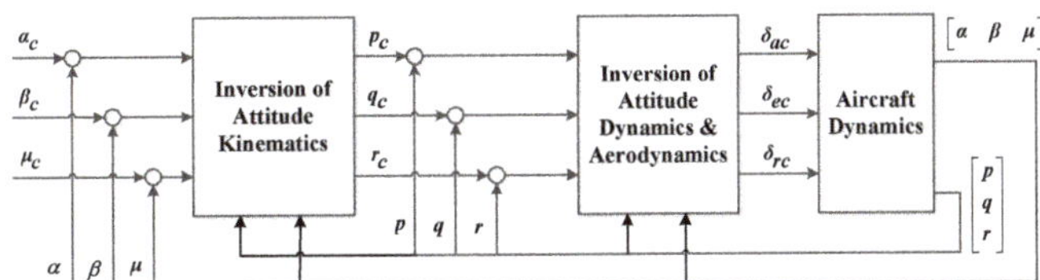

Figure 3. Configuration of the baseline nonlinear dynamic inversion (NDI) controller.

The inner loop is responsible for the inversion of the nonlinear aerodynamic database and attitude dynamics. For the application of NDI, the attitude dynamics are here rewritten in the state-space form:

$$\dot{x}_1 = f(x_1, x_2) + g(x_1, x_2)u \tag{16}$$

$$y_1 = x_1 \tag{17}$$

Recall the approach of feedback linearization, Equation (17) should be successively time differentiated until a relation between the input u and the output y_1 is found. After one time differentiation, Equation (17) yields

$$\dot{y}_1 = \dot{x}_1 = f(x_1, x_2) + g(x_1, x_2)u \tag{18}$$

Substituting the desired differentiated output with the pseudo-control input of Equation (19), the inversion control law of Equation (20) is obtained by inverting Equation (18), the resulting closed inner loop is expressed in Equation (21).

$$\dot{y}_1^d = B_1(x_1^c - x_1) \tag{19}$$

$$u = g^{-1}[B_1(x_1^c - x_1) - f] \tag{20}$$

$$\dot{x}_1 + B_1(x_1 - x_1^c) = 0 \tag{21}$$

where the superscript d represents the desired value, the superscript c represents the command from the outer loop as shown in Figure 3. The bandwidth matrix B_1, described in Equation (22), is responsible for stabilizing and improving the performance of the closed-loop system when the inversion or the model is not exact.

$$B_1 = \begin{bmatrix} k_p & 0 & 0 \\ 0 & k_q & 0 \\ 0 & 0 & k_r \end{bmatrix} \tag{22}$$

As the steady-state error in the inner loop will be solved by inserting an integrator in the outer loop, B_1 is only composed of the proportional elements. The gains of k_p, k_q, k_r are tuned to achieve rapid step response without exciting structural modes or exceeding the bandwidth limitations of the actuators [46].

The outer loop performs the inversion of the attitude kinematics equations, which are rewritten as:

$$\dot{x}_2 = h(x_1, x_2) \tag{23}$$

$$y_2 = x_2 \tag{24}$$

Similar to the inner loop, the output of Equation (24) is not immediately invertible and should be time differentiated once.

$$\dot{y}_2 = \dot{x}_2 = \begin{bmatrix} -\tan\beta\cos\alpha & 1 & -\tan\beta\sin\alpha \\ \sin\alpha & 0 & -\cos\alpha \\ \sec\beta\cos\alpha & 0 & \sec\beta\sin\alpha \end{bmatrix} x_1 \tag{25}$$

$$\dot{y}_2^d = B_2(x_2^c - x_2) \tag{26}$$

The outer loop control law is computed by substituting Equation (26) into Equation (25) and then inverting it.

$$x_1^c = \begin{bmatrix} -\tan\beta\cos\alpha & 1 & -\tan\beta\sin\alpha \\ \sin\alpha & 0 & -\cos\alpha \\ \sec\beta\cos\alpha & 0 & \sec\beta\sin\alpha \end{bmatrix}^{-1} B_2(x_2^c - x_2) \tag{27}$$

resulting in the closed outer loop

$$\dot{x}_2 + B_2(x_2 - x_2^c) = 0 \tag{28}$$

To improve the steady-state performance in the presence of modeling uncertainties, proportional-integral elements (NDI-PI) are used in the bandwidth matrix of B_2.

$$B_2 = \begin{bmatrix} k_{P_\alpha} + \frac{k_{I_\alpha}}{s} & 0 & 0 \\ 0 & k_{P_\beta} + \frac{k_{I_\beta}}{s} & 0 \\ 0 & 0 & k_{P_\mu} + \frac{k_{I_\mu}}{s} \end{bmatrix} \tag{29}$$

The tuning of $K_{P_{idx}}$ and $K_{I_{idx}}$, $idx = \alpha, \beta, \mu$ aims for best attitude position tracking performance, while avoiding actuator saturation.

3.3. Disturbance Observer Enhancement

In the case of the tailless aircraft model mounted on the 3-DOF rig in the wind tunnel, disturbances do exist as discussed in Equation (12), hence the baseline NDI control law is not necessary to result in a linear and time-invariant closed-loop system anymore. In particular, for the aforementioned inner loop, the closed-loop system of Equation (21) should be reformulated as Equation (31), when adding the mismatched disturbance d_1 into Equation (16).

$$\dot{x}_1 = f(x_1, x_2) + g(x_1, x_2)u + d_1 \tag{30}$$

$$\dot{x}_1 + B_1(x_1 - x_1^c) - d_1 = 0 \tag{31}$$

where d_1 is a lumped disturbance for the overall effect of modeling uncertainties and external disturbances as discussed in Section 2.

From a practical point of view, the lumped disturbance is continuous and bounded, thus satisfying Assumption 1.

Assumption 1. [47] The lumped disturbance and its derivative are bounded and satisfy $\|d_1\| \le K_1, \|\dot{d}_1\| \le K_2$, where $K_1 > 0$ and $K_2 > 0$ are specific constants.

The disturbance rejection can be reached by adding integral terms to the bandwidth matrix B_1 in Equation (22), however, this approach can remove the effect of the nearly constant disturbances in the steady-state but comes at the price of nominal performance degradation [27,48,49]. Instead, a nonlinear disturbance observer is designed and patched into the baseline NDI control law of the inner loop as shown in Figure 4. The NDI-DO controller is expected to reject the disturbance while preserving the nominal control performance.

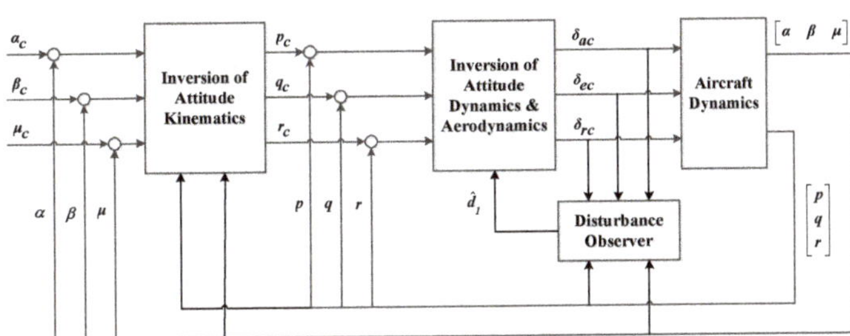

Figure 4. Disturbance observer-enhanced NDI controller.

As reported in ref. [47,50], the nonlinear disturbance observer is given by

$$\begin{cases} \dot{z}_{d_1} = -l(x_1)\left[\lambda(x_1) + z_{d_1} + f(x_1, x_2) + g(x_1, x_2)u\right] \\ \hat{d}_1 = z_{d_1} + \lambda(x_1) \end{cases} \tag{32}$$

where $\hat{d}_1$ is the estimation of d_1, z_{d_1} is the internal state vector of the nonlinear observer, and $\lambda(x_1)$ is a nonlinear function to be designed to satisfy $l(x_1) = \partial\lambda(x_1)/\partial x_1$.

After the lumped disturbance is estimated by the observer of Equation (32), the disturbance rejection NDI controller yields

$$u = g^{-1}\left[B_1(x_1^c - x_1) - f - \hat{d}_1\right] \tag{33}$$

It can be readily derived from Equations (30), (32), and (33) that

$$\dot{e}_{d_1} = -l(x_1)e_{d_1} + \dot{d}_1 \tag{34}$$

$$\dot{e}_{x_1} = -B_1 e_{x_1} + e_{d_1} \tag{35}$$

where the estimation error is defined as $e_{d_1} = d_1 - \hat{d}_1$, and the tracking error is defined as $e_{x_1} = x_1 - x_1^c$.

Considering the stability of the estimation error and tracking error, the choice of $\lambda(x_1)$ or $l(x_1)$ is appropriate such that the unforced dynamics $\dot{e}_{d_1} = -l(x_1)e_{d_1}$ is asymptotically stable at $e_{d_1} = 0$ [27,47].

According to the ref. [25,30,32], a simple and practical choice of $\lambda(x_1)$ is

$$\lambda(x_1) = Lx_1 = \begin{bmatrix} l_p & 0 & 0 \\ 0 & l_q & 0 \\ 0 & 0 & l_r \end{bmatrix} x_1 \tag{36}$$

where l_p, l_q, and l_r are positive constants.

In this case, the resulting gain matrix $-l(x_1) = -L$ is Hurwitz, and the estimation error dynamics can be expanded as

$$e_{d_1} = e^{-(t-t_0)L}e_{d_{10}} + \int_{t_0}^{t} e^{-(t-\tau)L}\dot{d}_1(\tau)d\tau \tag{37}$$

with $e_{d_{1,0}} = e_{d_1}(t_0)$ as the initial estimation error.

The solution of Equation (37) can be estimated using the bound of $\|e^{-(t-t_0)L}\| \leq \kappa_1 e^{-\kappa_2(t-t_0)}$ from the Theorem 4.11 of ref. [51].

$$
\begin{aligned}
\|e_{d_1}\| &\leq \|e^{-(t-t_0)L}\|\|e_{d_{1,0}}\| + \int_{t_0}^{t} \|e^{-(t-\tau)L}\| \sup_{t_0 \leq \tau \leq t} \|\dot{d}_1(\tau)\| \, d\tau \\
&\leq \kappa_1 e^{-\kappa_2(t-t_0)}\|e_{d_{1,0}}\| + \int_{t_0}^{t} \kappa_1 e^{-\kappa_2(t-t_0)} d\tau \sup_{t \geq t_0}\|\dot{d}_1(t)\| \\
&\leq \kappa_1 e^{-\kappa_2(t-t_0)}\|e_{d_{1,0}}\| + \frac{\kappa_1}{\kappa_2}\sup_{t \geq t_0}\|\dot{d}_1(t)\|
\end{aligned}
\tag{38}
$$

Thus, the ultimate bound of the estimation error is given by

$$
\lim_{t \to +\infty} \sup \|e_{d_1}\| \leq \frac{\kappa_1}{\kappa_2}\sup_{t \geq t_0}\|\dot{d}_1(t)\|
\tag{39}
$$

where κ_1 and κ_2 are positive constants decided by the Hurwitz matrix $-L$.

Under Assumption 1, it is noted from Equation (38) that the zero-input response decays to zero exponentially fast, while the zero-state response is bounded for every bounded input [27,47]. Additionally, Equation (39) shows that the estimation error can be finally regulated into an arbitrarily small neighborhood of zero with sufficiently small κ_1/κ_2. To do this, positive and sufficiently large elements of the gain matrix in Equation (36) are required to achieve faster observer dynamics than that of the disturbance [25,27].

Similarly, recalling the Hurwitz property of $-B_1$, the tracking error can be expressed as

$$
\begin{aligned}
\|e_{x_1}\| &\leq \|e^{-(t-t_0)B_1}\|\|e_{x_{1,0}}\| + \int_{t_0}^{t} \|e^{-(t-\tau)B_1}\| \sup_{t_0 \leq \tau \leq t} \|e_{d_1}(\tau)\| \, d\tau \\
&\leq \kappa_3 e^{-\kappa_4(t-t_0)}\|e_{x_{1,0}}\| + \int_{t_0}^{t} \kappa_3 e^{-\kappa_4(t-t_0)} d\tau \sup_{t \geq t_0}\|e_{d_1}(t)\| \\
&\leq \kappa_3 e^{-\kappa_4(t-t_0)}\|e_{x_{1,0}}\| + \frac{\kappa_3}{\kappa_4}\sup_{t \geq t_0}\|e_{d_1}(t)\|
\end{aligned}
\tag{40}
$$

From this, it follows that:

$$
\lim_{t \to +\infty} \sup \|e_{x_1}\| \leq \frac{\kappa_1 \kappa_3}{\kappa_2 \kappa_4}\sup_{t \geq t_0}\|\dot{d}_1(t)\|
\tag{41}
$$

where $e_{x_{1,0}} = e_{x_1}(t_0)$ denotes the initial value of tracking error and κ_3, κ_4 are positive constants decided by $-B_1$.

The effects of the lumped disturbance on the tracking error is bounded under Assumption 1 and can be mitigated with sufficiently small $\kappa_1\kappa_3/\kappa_2\kappa_4$. Namely, it is beneficial to reject the disturbance with larger gains in both the bandwidth matrix B_1 and the gain matrix L.

3.4. Anti-Windup Modification

For the tailless aircraft model, the actuators will saturate for surface commands out of the deflection ranges listed in Table 2.

Table 2. Deflection range of the surfaces on the tailless aircraft model.

Surface	Notation	Position Bound	Rate Limit
Aileron/Elevator	δ_a, δ_e	$\pm 30°$	$\pm 200°/s$
Rudder	δ_r	$\pm 40°$	$\pm 200°/s$

In the presence of actuator saturation, the system of Equation (30) takes the form:

$$
\dot{x}_1 = f(x_1, x_2) + g(x_1, x_2)sat(u) + d_1
\tag{42}
$$

$$sat(u) = \begin{bmatrix} sat_{\delta_a}(u_{\delta_a}) \\ sat_{\delta_e}(u_{\delta_e}) \\ sat_{\delta_r}(u_{\delta_r}) \end{bmatrix} \tag{43}$$

where sat_j is the saturation operator defined as following:

$$sat_j(u_j) = \begin{cases} u_j^{max}, & u_j \geq u_j^{max} \\ u_j, & u_i^{min} \leq u_j \leq u_j^{max} \\ u_j^{min}, & u_j \leq u_j^{min} \end{cases} \tag{44}$$

When saturation is active, the controller of Equation (33) exhibits windup, which could lead to significant performance degradation and even instability of the closed-loop system. To attenuate the effect of windup, an anti-windup modification is made to the disturbance observer of Equation (32) referring to ref. [25].

$$\begin{cases} \dot{z}_{d_1} = -l(x_1)\big[\lambda(x_1) + z_{d_1} + f(x_1, x_2) + g(x_1, x_2)u\big] \\ \qquad + K_{sat}[u - sat(u)] \\ \hat{d}_1 = z_{d_1} + \lambda(x_1) \end{cases} \tag{45}$$

where the term of K_{sat} is an anti-windup gain matrix to be designed.

Thus, using the modified observer of Equation (45) for estimation of the lumped disturbance in Equation (42), we have:

$$\dot{e}_{d_1} = -Le_{d_1} + \dot{d}_1 + [l(x_1)g(x_1, x_2) - K_{sat}][u - sat(u)] \tag{46}$$

The anti-windup gain matrix K_{sat} can be chosen as the following so that the effect of actuator saturation on the disturbance observer will be removed.

$$K_{sat} = l(x_1)g(x_1, x_2) \tag{47}$$

When Equation (47) is substituted into Equation (45), the modified disturbance observer yields

$$\begin{cases} \dot{z}_{d_1} = -l(x_1)\big[\lambda(x_1) + z_{d_1} + f(x_1, x_2) \\ \qquad + g(x_1, x_2)sat(u)\big] \\ \hat{d}_1 = z_{d_1} + \lambda(x_1) \end{cases} \tag{48}$$

The architecture of the closed-loop system with the actuator saturation and modified observer (NDI-AW) is as presented in Figure 5. Note that the anti-windup modified disturbance observer is driven by the actual surface deflections rather than the surface commands. Combining the plant dynamics of Equation (42), the inner-loop NDI controller of Equation (33) and the anti-windup modified disturbance observer of Equation (48), it is readily reached that the closed-loop tracking error takes the same form as Equation (35). Hence, the effect of actuator saturation on the closed-loop tracking performance is also removed.

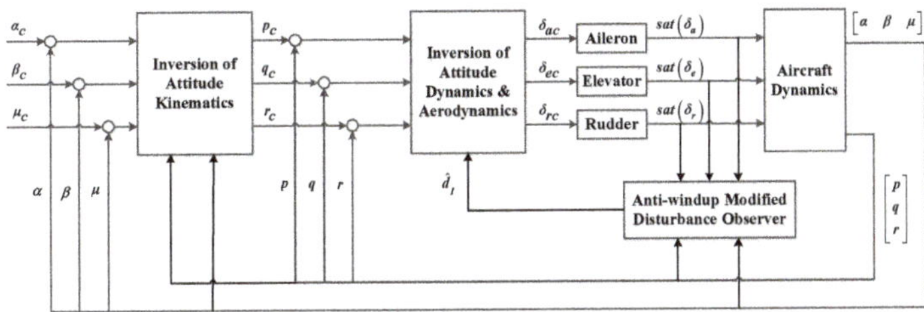

Figure 5. Anti-windup modification of the disturbance observer-enhanced NDI controller.

4. Simulation Study

This section describes the simulation testing of the three control schemes in attitude reference tracking, including the NDI-PI, the NDI-DO, and the NDI-AW. During the simulations, the aircraft model in the wind tunnel as detailed in Section 2 is augmented with second-order actuator dynamics for each surface, considering the deflection saturation in Table 2 and the time delay of 10 ms. The controller and observer gains are given in Table 3, which have been tuned to ensure tracking and robust performance in the presence of significant disturbances as listed in Table 4. For the sake of comparability, the nominal and robust simulations were all performed with the same step inputs on α_c and μ_c while minimizing the sideslip angle ($\beta_c = 0$) at the airspeed of 30 m/s. In contrast, step inputs were also applied to the sideslip angle in the anti-windup simulation.

Table 3. Controller and observer gains.

Coefficient	NDI-PI	NDI-DO	NDI-AW
K_p, K_q, K_r	20.0, 20.0, 10.0	10.0, 10.0, 5.0	10.0, 10.0, 5.0
$K_{P_\alpha}, K_{P_\beta}, K_{P_\mu}$	2.0, 2.0, 2.0	2.0, 2.0, 2.0	2.0, 2.0, 2.0
$K_{I_\alpha}, K_{I_\beta}, K_{I_\mu}$	0.1, 0.1, 0.1	0.2, 0.2, 0.2	0.2, 0.2, 0.2
l_p, l_q, l_r	—	15.0, 15.0, 15.0,	15.0, 15.0, 15.0

Table 4. Distributions for parameters affecting the robustness.

Parameter	Distribution	Mean	Standard Deviation
Noise of α, β, μ	normal	0	$2.0°$
Noise of p, q, r	normal	0	$0.4°/s$
Bias of $C_{i0}, \Delta C_{i_j}, \hat{C}_{i_k}$	normal	0	30%
Mismatch of CG	normal	0	5 mm

4.1. Comparison of the Nominal Performance

The nominal response of the three controllers to the tracking signal is given in Figure 6. As shown in Figure 6a, the reference signal of α_c was designed as a "square" with an amplitude of 5° and a width of 20 s. Each of the controller tracks the reference command well with a rise time of 1.0 s and an overshoot of 2%. As presented in Figure 6b, the reference signal of μ_c was a "doublet" with an amplitude of 5° and a width of 5 s. The rise time of the three controllers was around 0.8 s. However, there was an overshoot of 4% for the NDI-PI controller, while the other two controllers experienced no overshoot. In Figure 6c, it is observed that β tracking error was always smaller than 0.6°, which was the coupling response to the bank maneuvers. Moreover, the estimated lumped disturbances are shown in Figure 6d. Though there were notable estimation mismatches between the NDI-DO and the NDI-AW, the closed-loop performance under the two schemes were close to each other.

4.2. Comparison of the Robust Performance

The response of the tailless aircraft model controlled using the NDI-PI, NDI-DO, and NDI-AW with the sensor noise is shown in Figure 7. For comparison, the nominal response of the NDI-DO was also plotted with blue dotted lines. It turns out that the noise was not blown up for the three schemes, i.e., the steady responses of α, μ, and β were remarkably influenced by the sensor noise. However, it is to be mentioned that the disturbance observers succeeded to capture the disturbances induced by the bank maneuver as shown in Figure 7d.

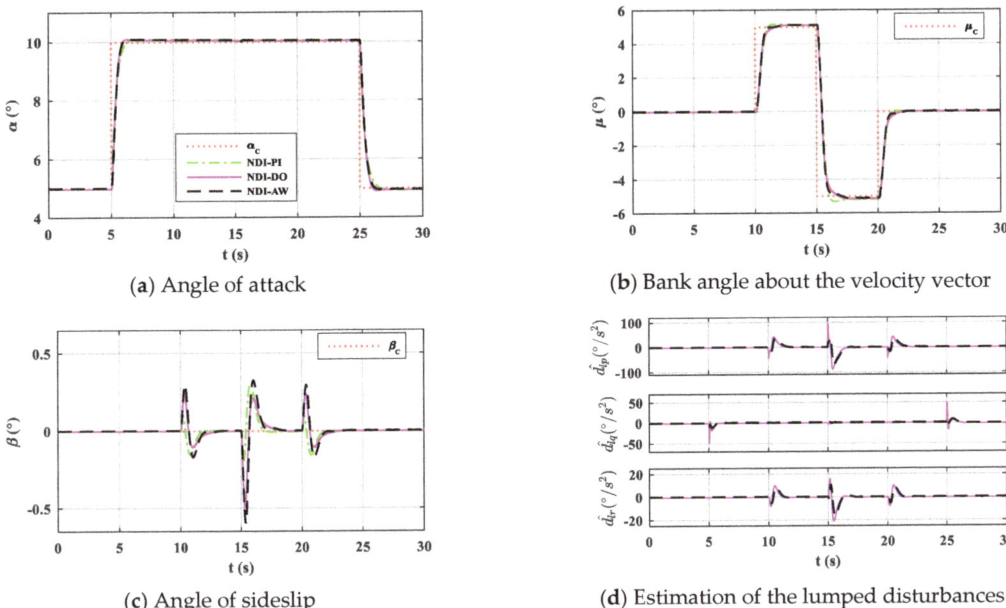

(**a**) Angle of attack

(**b**) Bank angle about the velocity vector

(**c**) Angle of sideslip

(**d**) Estimation of the lumped disturbances

Figure 6. The nominal response of the tailless aircraft model controlled using NDI-proportional-integral elements (PI), nonlinear disturbance operator based NDI (NDI-DO), and NDI-anti-windup (AW).

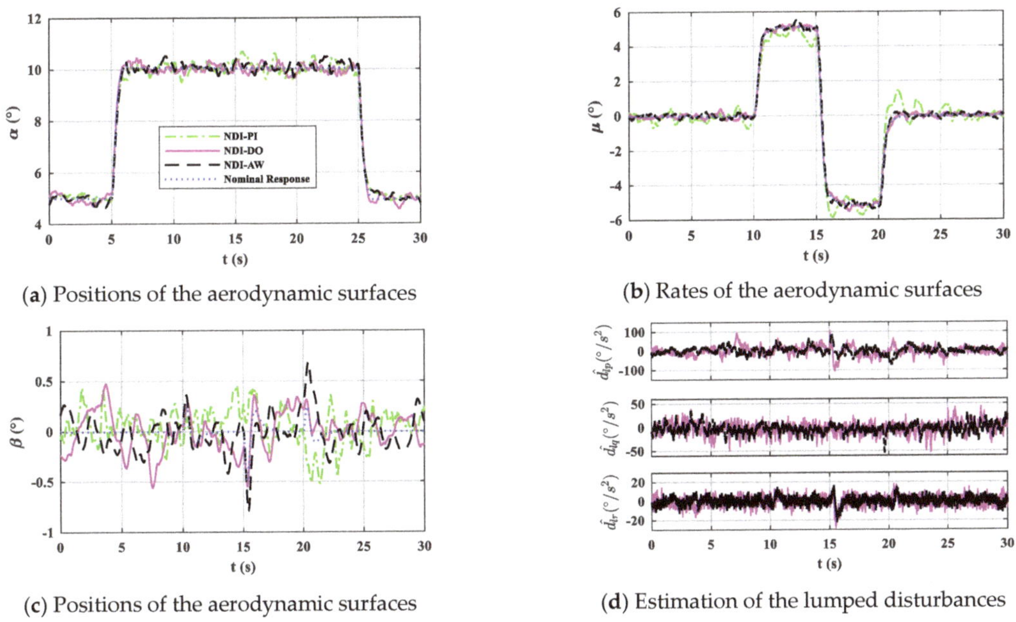

(**a**) Positions of the aerodynamic surfaces

(**b**) Rates of the aerodynamic surfaces

(**c**) Positions of the aerodynamic surfaces

(**d**) Estimation of the lumped disturbances

Figure 7. Response of the tailless aircraft model controlled using NDI-PI, NDI-DO, and NDI-AW with the sensor noise.

Figure 8 presents the robust simulation testing with the aerodynamic bias and the CG mismatch following the distributions in Table 4. Responses of five realizations in the distribution space are plotted for comparison with the nominal response of the NDI-DO controller.

It is noted that the NDI-PI controller suffers from the proposed disturbances and the tracking performance is unsatisfactory. As discussed in ref [36], it is the CG mismatch, especially along the aircraft's longitudinal axis, which will bring remarkable external disturbances to the aircraft's flight dynamics. As a result, the step responses of NDI-PI are scattered as shown in Figure 8a,b. In the case of α, remarkable response undershoot or overshoot is observed for each realization ranged from about -34% to $+68\%$ of the reference step signal, and the integrators in the outer loop controller of Equation (29) tend to eliminate the steady error gradually. In the case of μ, a significant tracking error is caused not only by the bank maneuver but also by the coupling effect of the pitch maneuver. In the case of β, the performance is satisfactory with a slightly increased magnitude of coupling error.

Compared with the NDI-PI controller, the NDI-DO and NDI-AW controllers are not sensitive to the aerodynamic biases or the CG mismatch. The α and μ responses are recovered close to that of the nominal case, except that the overshoots and damping deteriorate marginally, and the coupling influence of μ maneuver to α and β increases slightly. It is the disturbance observers of Equations (32) and (48) that estimate the aerodynamic biases and the CG mismatch as a set of lumped disturbances, respectively, as shown in Figure 8d. Additionally, then, the estimated lumped disturbances are substituted into Equation (33) for dynamic inversion compensation. As a result, the disturbance rejection abilities of NDI-DO and NDI-AW have been improved significantly. Besides, the notable coupling error is observed in μ and β at about 6 s and 26 s, when the aircraft model nose up or nose down rapidly. As is well known, the magnitude of aerodynamic forces applied to the aircraft varies a lot along with the angle of attack. In the presence of the CG mismatch, significant moment shifts are induced as shown in Figure 8d. Consequently, the tracking performance of the NDI-DO and the NDI-AW controllers decays along with the rapid fluctuations of the lumped disturbances.

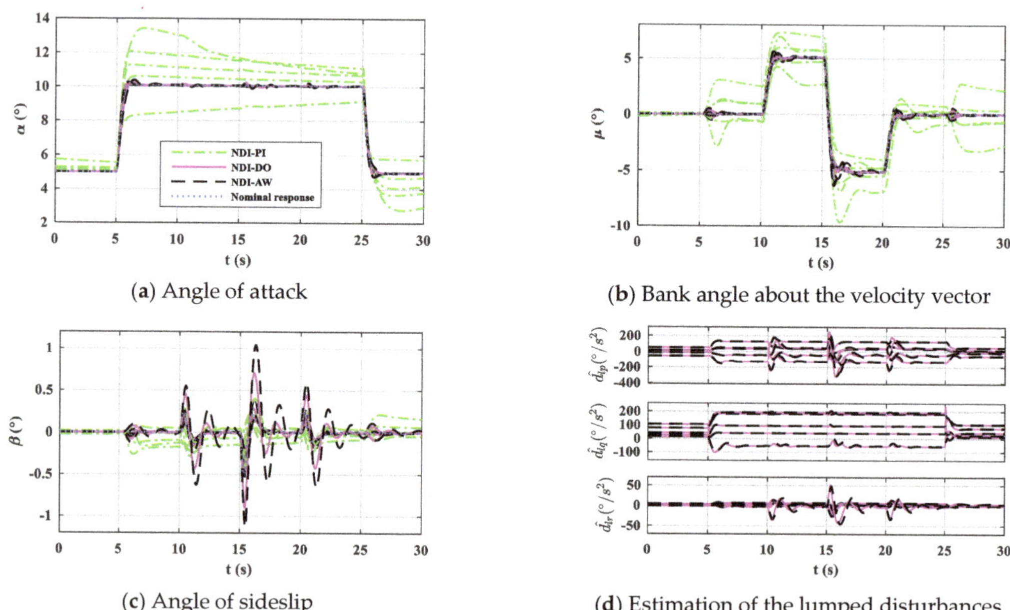

Figure 8. Response of the tailless aircraft model controlled using NDI-PI, NDI-DO, and NDI-AW with the aerodynamic bias and the center of gravity (CG) mismatch.

4.3. Investigation of the Anti-Windup Performance

In this part, aggressive maneuvers for the longitudinal, lateral, and directional channels are performed with their respective square and doublet reference inputs. The amplitude of the maneuvers is chosen through a trial-and-error approach to ensure that the problem of actuator saturation is excited. For the sake of simplicity, the disturbances listed in Table 4 were turned off for this simulation.

The results are shown in Figure 9 with the corresponding responses of the aircraft model controlled by the three schemes. It turns out that the both the NDI-PI and NDI-DO schemes suffered from the large amplitude maneuvers, while the NDI-AW scheme shows significantly improved performance.

The first remarkable response mismatch of the three schemes is observed during the time from about 10 to 15 s, i.e., the tracking performance of the NDI-PI scheme deteriorated in terms of furious coupling, overshoot, oscillation and steady-error. However, the tracking performances of the other two schemes were satisfactory and close to each other. Such an outcome was in agreement with the prospective effect of the disturbance observer. On one hand, the NDI-DO and the NDI-AW controllers estimated the lumped disturbances using Equations (32) and (48), respectively, which were then compensated using the inner-loop dynamic inversion controller of Equation (33). As a result, the influence of the lumped disturbances to the closed-loop system was mostly rejected by the disturbance observer augmented controllers. On the other hand, though transient saturation was observed for both the NDI-DO and the NDI-AW schemes, the problem of windup was not excited sufficiently to cause the closed-loop performance decay.

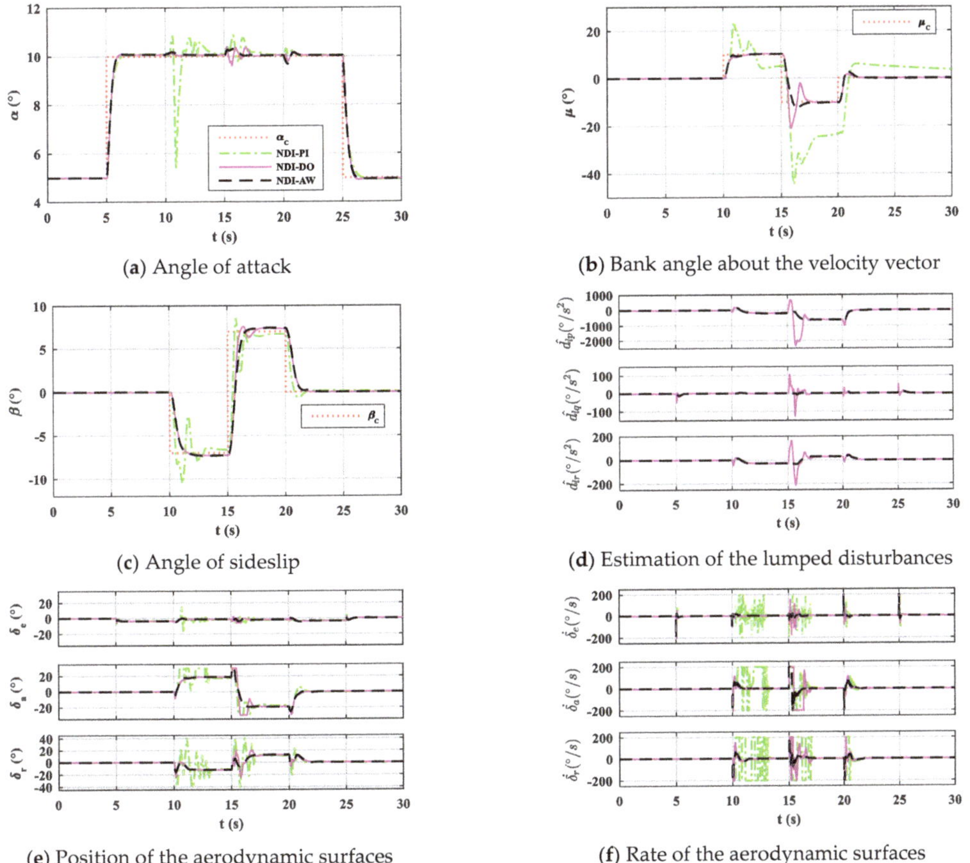

(a) Angle of attack

(b) Bank angle about the velocity vector

(c) Angle of sideslip

(d) Estimation of the lumped disturbances

(e) Position of the aerodynamic surfaces

(f) Rate of the aerodynamic surfaces

Figure 9. Response of the tailless aircraft model controlled using NDI-PI, NDI-DO, and NDI-AW with the actuator saturation.

During the bank and sideslip maneuver from 15 to 20 s, the performance of not only the NDI-PI but also the NDI-DO deteriorated significantly, while the response of the NDI-AW was still satisfactory with marginal overshoot in μ and coupling error in α. Regarding the performance maintenance of the NDI-AW scheme in the presence of actuator saturation, the effect of the anti-windup modified observer was investigated. As soon as the deflections of the aerodynamic surfaces saturate, the anti-windup modification of Equation (48) in the NDI-AW scheme became active, which took the actuator induced delay and error into account. Hence, the deflection saturations induced nonlinear disturbance is estimated and compensated by the NDI-AW scheme, whereas this is not the fact for the NDI-DO scheme. As a result, the estimated lumped disturbance of the NDI-AW in Figure 9d was of much smaller magnitude than that of the NDI-DO. Furthermore, according to Equation (39), the estimation error of the disturbance observer was related to the time derivative of the disturbances. Since the lumped disturbance of the NDI-DO consists of the nonlinear, noncontinuous deflection saturations, the resulting estimation error blows up. This explains the performance decay of the NDI-DO scheme.

5. Experimental Results

A flight test campaign was conducted in the wind tunnel using the NDI-PI and NDI-DO schemes. Due to the difficulties to sense the actual deflections of the surfaces, the NDI-AW scheme was not implemented in the test. As shown in Figure 10, the same reference inputs as that of the nominal and robust simulations in Section 4 were applied to the aircraft model for multiple times at the wind speed of 20 m/s, 25 m/s, and 30 m/s, respectively, to obtain repeatable responses of the aircraft model. The controller was selected and activated via the signal of "NoCtrl" from a switch button of the pilot stick, which is a discrete variable corresponding to the case of open-loop (OFF) with "0", NDI-PI with "1", and NDI-DO with "2". For the sake of controller initialization, switching between the NDI-PI and the NDI-DO was always interpolated with "OFF" for about 1 s. One can note that the attitude responses of the aircraft model under the NDI-PI and NDI-DO were reasonable, except that notable perturbations were observed at around 290 s, 630 s due to step increase of the wind speed, and at about 174 s, 210 s, 347 s, 415 s, and 792 s due to outliers of the IMU sensor.

The experimental data under the NDI-PI and NDI-DO at the wind speed of 30 m/s are presented in Figure 11 in comparison to the nominal simulation responses of NDI-DO. For the sake of comparability, the time axis of the experimental data is shifted to be coincident with the simulation, i.e., from 0 to 30 s. One can see that experimental responses agreed well with the simulation, though high-frequency noise has been introduced into the attitude by the sensors. In general, the experimental responses were satisfactory with an overshoot of about 11% in α under the NDI-DO, an overshoot of about 14% and a slight decay of damping in μ under the NDI-PI. Performances of both the NDI-PI and the NDI-DO schemes deteriorated due to the disturbances in the experimental setup. As shown in Figure 11d, the estimated lumped disturbances of NDI-DO in the test shifted and deviated remarkably from that of NDI-DO in the simulation. According to the results and analysis of the robust simulation in Section 4, it is inferred that there were marginal mismatches between the CG of the aircraft model and the rotational center of the support rig. Besides, moderate disturbances may result from the bias and noise of the sensors, the aerodynamic influence of the support rig, the error of aircraft mass and inertia, the fluctuation of wind speed, etc. Compared to the NDI-PI scheme, the disturbance observer augmented scheme of NDI-DO decreases the coupling error in α and β during the bank maneuver.

In summary, the experimental results show that both the NDI-PI and the NDI-DO schemes worked well to stabilize and control the attitude of the aircraft model in the wind tunnel tests. The NDI-DO scheme demonstrated slightly improved tracking performance than the NDI-PI scheme in the presence of the disturbances in the experimental setup.

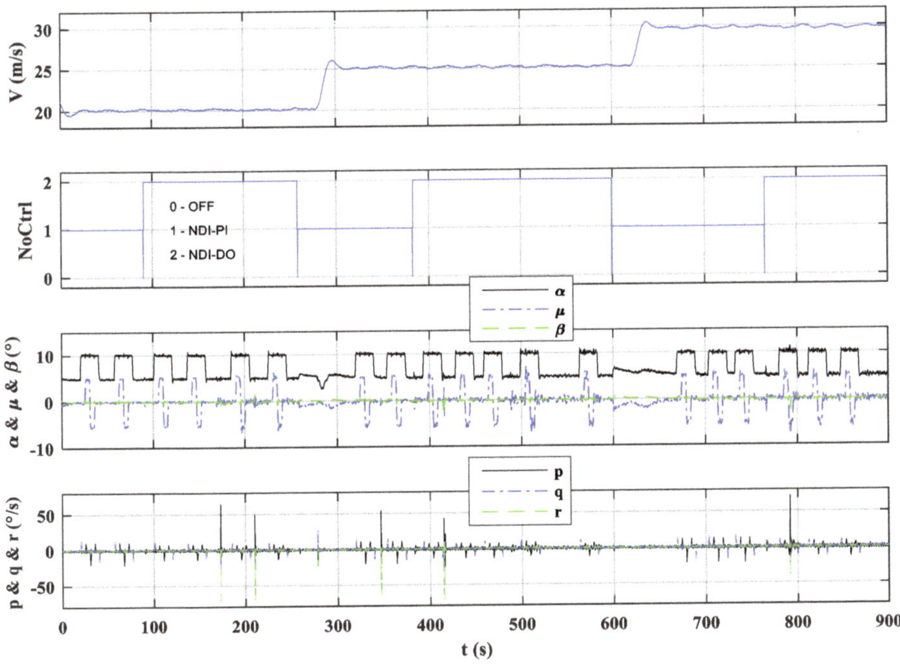

Figure 10. Experimental results of the tailless aircraft model controlled using NDI-PI and NDI-DO in the wind tunnel flight tests.

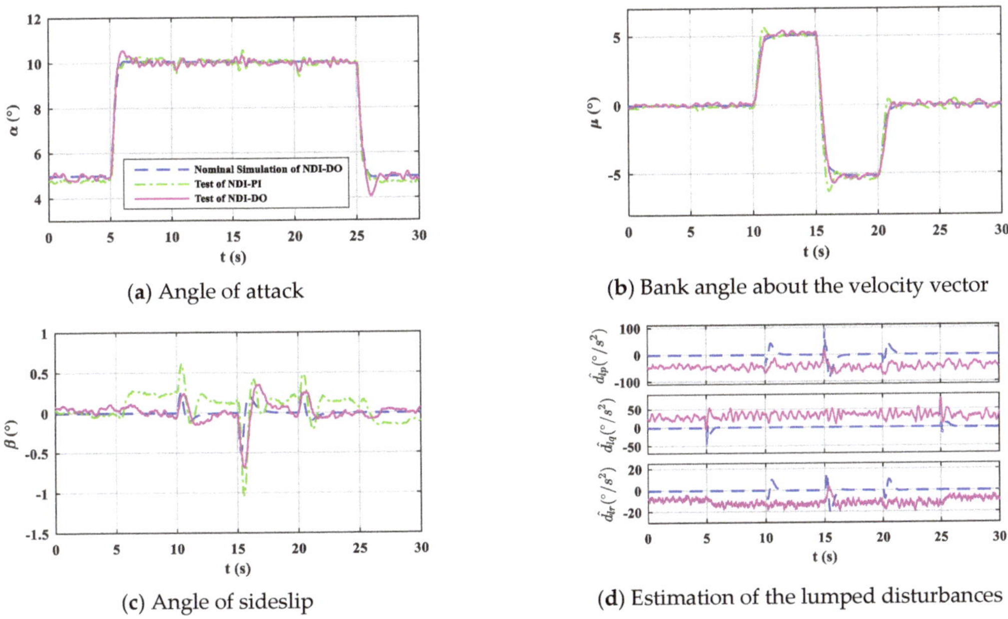

(**a**) Angle of attack

(**b**) Bank angle about the velocity vector

(**c**) Angle of sideslip

(**d**) Estimation of the lumped disturbances

Figure 11. Experimental responses of the tailless aircraft model controlled using NDI-PI and NDI-DO in the wind tunnel flight tests in contrast with the nominal simulation results of NDI-DO.

6. Conclusions

In this paper, the dynamic inversion based disturbance rejection control schemes were designed and subsequent flight tests in the wind tunnel were carried out for a tailless aircraft model. An overview of the aircraft model and the experimental setup was provided. A comprehensive nonlinear mathematical model related to the flight dynamics of the aircraft model supported with a 3-DOF rig in the wind tunnel was derived and then rewritten in the form of a nonlinear affine system. A baseline NDI controller was designed to stabilize and control the aircraft attitude, which was further augmented with a disturbance observer to reject the lumped disturbances. In the presence of actuator saturation, an anti-windup modification was made to the disturbance observer. The simulation results show that the robustness of the disturbance observer augmented controllers was superior to the baseline NDI controller, and the anti-windup modified disturbance observer helped to recover the control performance from the actuator saturation. Finally, flight tests were successfully conducted in the wind tunnel setup showing that the experimental results agreed well with the simulations under the control of the NDI-DO scheme, which demonstrated superior tracking and robust performance than the NDI-PI scheme. However, the NDI-AW controller was not implemented and tested due to the absence of sensors for the actual surface deflections.

To the authors' knowledge, this is the first realistic tailless aircraft model application and real-time implementation of the NDI-DO scheme to validate the effectiveness of the disturbance rejection NDI control method in the wind tunnel. It is shown that the NDI-DO scheme performed better than the NDI-PI scheme in the presence of high-level disturbances. The NDI-AW scheme shows the prospective closed-loop performance recovering from the actuator saturations, which will be implemented and validated via the wind-tunnel flight tests in the future.

Author Contributions: Supervision, H.M. and O.S.; Writing—Original draft, B.N.; Writing—Review and editing, Z.L., T.G. and L.F. All authors have read and agreed to the published version of the manuscript.

Funding: This research received no external funding.

Institutional Review Board Statement: Not applicable.

Informed Consent Statement: Not applicable.

Data Availability Statement: The data presented in this study are available on request from the corresponding author.

Conflicts of Interest: The authors declare no conflict of interest.

References

1. Colgren, R.; Loschke, R. Effective design of highly maneuverable tailless aircraft. *J. Aircr.* **2008**, *45*, 1441–1449. [CrossRef]
2. Friehmelt, H. Thrust vectoring and tailless aircraft design-Review and outlook. In Proceedings of the 21st Atmospheric Flight Mechanics Conference, San Diego, CA, USA, 29–31 July 1996; pp. 420–430.
3. Bowlus, J.; Multhopp, D.; Banda, S.; Bowlus, J.; Multhopp, D.; Banda, S. Challenges and opportunities in tailless aircraft stability and control. In *Guidance, Navigation, and Control Conference*; American Institute of Aeronautics and Astronautics: Boston, MA, USA, 1997; pp. 1713–1718.
4. Wise, K. Fighter Aircraft Control Challenges and Technology Transition. In *Systems and Control in the Twenty-First Century*; Birkhäuser: Boston, MA, USA, 1997; pp. 417–434. ISBN 978-1-4612-8662-2.
5. Stenfelt, G.; Ringertz, U. Yaw control of a tailless aircraft configuration. *J. Aircr.* **2010**, *47*, 1807–1811. [CrossRef]
6. Stenfelt, G.; Ringertz, U. Yaw departure and recovery of a tailless aircraft configuration. *J. Aircr.* **2013**, *50*, 311–315. [CrossRef]
7. Wells, S.R.; Hess, R.A. Multi-Input/Multi-Output Sliding Mode Control for a Tailless Fighter Aircraft. *J. Guid. Control. Dyn.* **2003**, *26*, 463–473. [CrossRef]
8. Ngo, A.; Reigelsperger, W.; Banda, S.; Bessolo, J. Multivariable control law design for a tailless airplane. In Proceedings of the Guidance, Navigation, and Control Conference, San Diego, CA, USA, 29–31 July 1996; pp. 1–11.
9. Buffington, J. *Modular Control. Law Design for the Innovative Control. Effectors (ICE) Tailless Fighter Aircraft Configuration 101-3*; AFRL-VA-WP-TP-1999-3057; U.S. Air Force Research Lab.: Wright-Patterson Air Force Base, OH, USA, June 1999.
10. Wise, K.; Sedwick, J.; Ikeda, Y. *Nonlinear Control. of Fighter Aircraft*; F49620-96-C-0011; The Boeing Company: St. Louis, MO, USA, June 1999.

11. Jackson, E.B.; Buttrill, C.W. *Control. Laws for a Wind Tunnel Free-Flight Study of a Blended-Wing-Body Aircraft*; TM–2006–214501; NASA: Hampton, VA, USA, 2006.

12. Cook, M.V.; Castro, H.V. The longitudinal flying qualities of a blended-wing-body civil transport aircraft. *Aeronaut. J.* **2004**, *108*, 75–84. [CrossRef]

13. Nieto-Wire, C.; Sobel, K. Delta Operator Eigenstructure Assignment for Reconfigurable Control of a Tailless Aircraft. *J. Guid. Control. Dyn.* **2014**, *37*, 1824–1839. [CrossRef]

14. Calise, A.; Lee, S.; Sharma, M. Direct adaptive reconfigurable control of a tailless fighter aircraft. In Proceedings of the Guidance, Navigation, and Control Conference and Exhibit, Boston, MA, USA, 10–12 August 1998.

15. Brinker, J.; Wise, K. Reconfigurable flight control for a tailless advanced fighter aircraft. In Proceedings of the Guidance, Navigation, and Control Conference and Exhibit, Boston, MA, USA, 10–12 August 1998.

16. Calise, A.J.; Lee, S.; Sharma, M. Development of a Reconfigurable Flight Control Law for Tailless Aircraft. *J. Guid. Control. Dyn.* **2001**, *24*, 896–902. [CrossRef]

17. Brinker, J.S.; Wise, K.A. Flight Testing of Reconfigurable Control Law on the X-36 Tailless Aircraft. *J. Guid. Control. Dyn.* **2001**, *24*, 903–909. [CrossRef]

18. Jones, C.; Lowenberg, M.; Richardson, T. Dynamic State-Feedback Gain Scheduled Control of the ICE 101-TV. In Proceedings of the AIAA Guidance, Navigation, and Control Conference and Exhibit, Providence, RI, USA, 16–19 August 2004.

19. Huber, P. Control law design for tailless configurations and in-flight simulation using the X-31 aircraft. In Proceedings of the Guidance, Navigation, and Control Conference, Baltimore, MD, USA, 7–10 August 1995.

20. Balas, G.J. Flight Control Law Design: An Industry Perspective. *Eur. J. Control.* **2003**, *9*, 207–226. [CrossRef]

21. Enns, D.; Bugajski, D.; Hendrick, R.; Stein, G. Dynamic inversion: An evolving methodology for flight control design. *Int. J. Control.* **1994**, *59*, 71–91. [CrossRef]

22. Wang, X.; van Kampen, E.-J.; Chu, Q.; Lu, P. Stability Analysis for Incremental Nonlinear Dynamic Inversion Control. *J. Guid. Control. Dyn.* **2019**, *42*, 1116–1129. [CrossRef]

23. Safwat, E.; Zhang, W.; Mohsen, A. Design and Analysis of a Robust UAV Flight Guidance and Control System Based on a Modified Nonlinear Dynamic Inversion. *Appl. Sci.* **2019**, *9*, 3600. [CrossRef]

24. Vicroy, D.D.; Loeser, T.D.; Schütte, A. Static and forced-oscillation tests of a generic unmanned combat air vehicle. *J. Aircr.* **2012**, *49*, 1558–1583. [CrossRef]

25. Smith, J.; su, J.; Liu, C.; Chen, W.-H. Disturbance Observer Based Control with Anti-Windup Applied to a Small Fixed Wing UAV for Disturbance Rejection. *J. Intell. Robot. Syst.* **2017**, *88*, 329–346. [CrossRef]

26. Ding, S.; Chen, W.; Mei, K.; Murray-Smith, D.J. Disturbance Observer Design for Nonlinear Systems Represented by Input–Output Models. *IEEE Trans. Ind. Electron.* **2020**, *67*, 1222–1232. [CrossRef]

27. An, H.; Wu, Q. Disturbance rejection dynamic inverse control of air-breathing hypersonic vehicles. *Acta Astronaut.* **2018**, *151*, 348–356. [CrossRef]

28. Yang, J.; Zolotas, A.; Chen, W.-H.; Michail, K.; Li, S. Robust control of nonlinear MAGLEV suspension system with mismatched uncertainties via DOBC approach. *ISA Trans.* **2011**, *50*, 389–396. [CrossRef]

29. Chen, W.-H.; Ballance, D.J.; Gawthrop, P.J.; O'Reilly, J. A nonlinear disturbance observer for robotic manipulators. *IEEE Trans. Ind. Electron.* **2000**, *47*, 932–938. [CrossRef]

30. Chen, W.-H. Nonlinear Disturbance Observer-Enhanced Dynamic Inversion Control of Missiles. *J. Guid. Control. Dyn.* **2003**, *26*, 161–166. [CrossRef]

31. Sun, L.; Zheng, Z. Disturbance Observer-Based Robust Saturated Control for Spacecraft Proximity Maneuvers. *IEEE Trans. Control. Syst. Technol.* **2018**, *26*, 684–692. [CrossRef]

32. Chen, Z.; Sun, C.; Cen, F.; Nie, B.; Li, Q. Disturbance observer based dynamic inversion control for a wind tunnel free-flying test of a blended-wing-body aircraft. In Proceedings of the 2019 Chinese Automation Congress, Hangzhou, China, 22–24 November 2019.

33. Valasek, J.; Ito, D.; Ward, D. Robust dynamic inversion controller design and analysis for the X-38. In Proceedings of the AIAA Guidance, Navigation, and Control Conference and Exhibit, Montreal, QC, Canada, 6–9 August 2001.

34. Menon, P.P.; Lowenberg, M.; Herrmann, G.; Turner, M.C.; Bates, D.G.; Postlethwaite, I. Experimental Implementation of a Nonlinear Dynamic Inversion Controller with Antiwindup. *J. Guid. Control. Dyn.* **2013**, *36*, 1035–1046. [CrossRef]

35. Herrmann, G.; Menon, P.P.; Turner, M.C.; Bates, D.G.; Postlethwaite, I. Anti-windup synthesis for nonlinear dynamic inversion control schemes. *Int. J. Robust Nonlinear Control.* **2010**, *20*, 1465–1482. [CrossRef]

36. Nie, B.; Liu, Z.; Cen, F.; Liu, D.; Ma, H.; Sename, O. An Innovative Experimental Approach to Lateral-Directional Flying Quality Investigation for Tailless Aircraft. *IEEE Access* **2020**, *8*, 109543–109556. [CrossRef]

37. Chen, Z.; Li, Q.; Ju, X.; Cen, F. Barrier Lyapunov Function based Sliding Mode Control for BWB Aircraft with Mismatched Disturbances and Output Constraints. *IEEE Access* **2019**, 1. [CrossRef]

38. Stevens, B.L.; Lewis, F.L.; Johnson, E.N. *Aircraft Control and Simulation: Dynamics, Controls Design, and Autonomous Systems*, 3rd ed.; Wiley: Hoboken, NJ, USA, 2016.

39. Juliana, S.; Chu, Q.P.; Mulder, J.A.; van Baten, T.J. Flight Control of Atmospheric Re-entry Vehicle with Non-linear Dynamic Inversion. In Proceedings of the AIAA Guidance, Navigation, and Control Conference and Exhibit, Providence, RI, USA, 16–19 August 2004.

40. Snell, S.A.; Enns, D.F.; Garrard, W.L. Nonlinear inversion flight control for a supermaneuverable aircraft. *J. Guid. Control. Dyn.* **1992**, *15*, 976–984. [CrossRef]
41. Lane, S.H.; Stengel, R.F. Flight control design using non-linear inverse dynamics. *Automatica* **1988**, *24*, 471–483. [CrossRef]
42. Hedrick, J.K.; Gopalswamy, S. Nonlinear flight control design via sliding methods. *J. Guid. Control. Dyn.* **1990**, *13*, 850–858. [CrossRef]
43. Sieberling, S.; Chu, Q.P.; Mulder, J.A. Robust Flight Control Using Incremental Nonlinear Dynamic Inversion and Angular Acceleration Prediction. *J. Guid. Control Dyn.* **2010**, *33*, 1732–1742. [CrossRef]
44. Adams, R.J.; Buffington, J.M.; Banda, S.S. Design of nonlinear control laws for high-angle-of-attack flight. *J. Guid. Control Dyn.* **1994**, *17*, 737–746. [CrossRef]
45. Bugajski, D.J.; Enns, D.F. Nonlinear control law with application to high angle-of-attack flight. *J. Guid. Control Dyn.* **1992**, *15*, 761–767. [CrossRef]
46. da Costa, R.R.; Chu, Q.P.; Mulder, J.A. Reentry flight controller design using nonlinear dynamic inversion. *J. Spacecr. Rockets* **2003**, *40*, 64–71. [CrossRef]
47. Yang, J.; Li, S.; Chen, W.-H. *Nonlinear Disturbance Observer-Based Control for Multi-Input Multi-Output Nonlinear Systems Subject to Mismatching Condition*; Taylor & Francis: Abingdon, UK, 2012; Volume 85.
48. Rehman, O.U.; Fidan, B.; Petersen, I.R. Uncertainty modeling and robust minimax LQR control of multivariable nonlinear systems with application to hypersonic flight. *Asian J. Control.* **2012**, *14*, 1180–1193. [CrossRef]
49. Wang, Q.; Stengel, R.F. Robust Nonlinear Control of a Hypersonic Aircraft. *J. Guid. Control. Dyn.* **2000**, *23*, 577–585. [CrossRef]
50. Chen, W.H.; Ballance, D.J.; Gawthrop, P.J.; Gribble, J.J.; O'Reilly, J. Nonlinear PID predictive controller. *IEE Proc. Control. Theory Appl.* **1999**, *146*, 603–611. [CrossRef]
51. Khalil, H.K.; Grizzle, J.W. *Nonlinear Systems*, 3rd ed.; Prentice Hall: Upper Saddle River, NJ, USA, 2002; ISBN 0130673897.

MDPI

St. Alban-Anlage 66

4052 Basel

Switzerland

Tel. +41 61 683 77 34

Fax +41 61 302 89 18

www.mdpi.com

Applied Sciences Editorial Office

E-mail: applsci@mdpi.com

www.mdpi.com/journal/applsci